Soils and Food Security

ISSUES IN ENVIRONMENTAL SCIENCE AND TECHNOLOGY

TITLES IN THE SERIES:

1: Mining and its Environmental Impact
2: Waste Incineration and the Environment
3: Waste Treatment and Disposal
4: Volatile Organic Compounds in the Atmosphere
5: Agricultural Chemicals and the Environment
6: Chlorinated Organic Micropollutants
7: Contaminated Land and its Reclamation
8: Air Quality Management
9: Risk Assessment and Risk Management
10: Air Pollution and Health
11: Environmental Impact of Power Generation
12: Endocrine Disrupting Chemicals
13: Chemistry in the Marine Environment
14: Causes and Environmental Implications of Increased UV-B Radiation
15: Food Safety and Food Quality
16: Assessment and Reclamation of Contaminated Land
17: Global Environmental Change
18: Environmental and Health Impact of Solid Waste Management Activities
19: Sustainability and Environmental Impact of Renewable Energy Sources
20: Transport and the Environment
21: Sustainability in Agriculture
22: Chemicals in the Environment: Assessing and Managing Risk
23: Alternatives to Animal Testing
24: Nanotechnology
25: Biodiversity Under Threat
26: Environmental Forensics
27: Electronic Waste Management
28: Air Quality in Urban Environments
29: Carbon Capture
30: Ecosystem Services
31: Sustainable Water
32: Nuclear Power and the Environment
33: Marine Pollution and Human Health
34: Environmental Impacts of Modern Agriculture
35: Soils and Food Security

How to obtain future titles on publication

A subscription is available for this series. This will bring delivery of each new volume immediately on publication and also provide you with online access to each title via the Internet. For further information visit http://www.rsc.org/issues or write to the address below.

For further information please contact:
Sales and Customer Care, Royal Society of Chemistry, Thomas Graham House, Science Park, Milton Road, Cambridge, CB4 0WF, UK
Telephone: +44 (0)1223 432360, Fax: +44 (0)1223 426017, Email: booksales@rsc.org

ISSUES IN ENVIRONMENTAL SCIENCE AND TECHNOLOGY

EDITORS: R.E. HESTER AND R.M. HARRISON

35
Soils and Food Security

RSC Publishing

ISBN: 978-1-84973-426-4
ISSN 1350-7583

A catalogue record for this book is available from the British Library

Published by The Royal Society of Chemistry,
Thomas Graham House, Science Park, Milton Road,
Cambridge CB4 0WF, UK

Registered Charity Number 207890

For further information see our web site at www.rsc.org

Printed in the United Kingdom by CPI Group (UK) Ltd, Croydon, CR0 4YY, UK

Preface

The world's population reached 7 billion in 2011 and is expected to grow further to exceed 9 billion by 2050. Feeding this population depends on the thin layer of soil which covers the Earth's ice-free land surface, less than 40% of which is currently available to agriculture. The demand for food, feed and fibre is expected to increase by 70% by 2050, but around 60% of the soil resources needed have already been degraded by soil erosion and depletion of nutrients or are currently being used unsustainably. Throughout much of the less developed world, where the rate of population growth is highest (the current 1 billion population of Africa is expected to double by 2050), food production is dominated by low yields from unproductive soils. Cropping the land removes nutrients from the soil and the replacement levels of fertilisers required for sustainability are simply unaffordable by most African farmers. The resulting soil degradation and erosion has serious adverse consequences, not only for food production (850 million people currently suffer from undernourishment) but also for other important ecosystem services provided by soils (carbon and water storage, filtration and buffering, waste disposal, nutrient recycling and the support of genetic diversity).

Land and soil are limited and non-renewable resources. But, if nothing changes, by 2050 the equivalent of two planet Earths will be needed to sustain the population. With growing affluence in some parts of the developed world, increasing consumption of meat and dairy products and the use of biomass for energy and other industrial purposes create increasing demand for agricultural production. Meanwhile, increasing urbanisation (growth of cities and industrialisation), infrastructure growth and land-use changes (*e.g.* biofuels production), are shrinking the land base available for agriculture.

Soil functions as both a source and a sink for carbon dioxide and has a powerful influence on global climate. There is twice as much carbon in soil worldwide than in the whole of the Earth's atmosphere. It is well known that

Issues in Environmental Science and Technology No. 35
Soils and Food Security
Edited by R.E. Hester and R.M. Harrison

Published by the Royal Society of Chemistry, www.rsc.org

forestry has important influences on climate but less known that the potent greenhouse gases methane and nitrous oxide can be released from wetlands and permafrost soils as climate warming occurs, thus providing a major feedback mechanism for enhanced warming. These climate change effects, in turn, impact on agricultural productivity.

From these introductory observations it may reasonably be concluded that there is emerging an urgent need to combat erosion of the world's soils and to use them sustainably for food production if we are to avoid problems of severe and widespread malnutrition or even famine in the not-so-distant future. This book is an attempt to address this problem area. It begins with a chapter by Peter Gregory, Chief Executive of East Malling Research in Kent, UK, which surveys the challenges and opportunities for sustaining soil fertility while avoiding detrimental environmental consequences. Luca Montanarella of the EU Joint Research Centre in Ispra, Italy, then reviews the problem of preserving the capacity of global soils for food production. In Chapter 3, David Robinson, of the NERC Centre for Ecology and Hydrology in Bangor, Wales, in collaboration with colleagues from other UK institutions, discusses the concepts of soil natural capital and ecosystem service delivery. Next is a review by Alfred Hartemink and colleagues from the University of Wisconsin, USA, of the evaluation and reporting of soils in sustainable agriculture and food systems. The focus then shifts to sub-Saharan Africa (SSA) in Chapters 5 and 6. Fredrick Ayuke and his colleagues at the University of Nairobi, Kenya, and the World Agroforestry Centre in Nairobi, discuss agrobiodiversity and its potential use for enhancing soil health in the tropical soils of Africa. Then Bernard Vanlauwe of the International Institute of Tropical Agriculture in Nairobi examines the availability and management of soil organic matter in the context of Integrated Soil and Fertility Management in SSA. Chapter 7, written by Wendy Peterman and Dominique Bachelet of the Conservation Biology Institute in Corvallis, USA, addresses the key topic of climate change in the context of forest dynamics. In Chapter 8, David Manning of Newcastle University, UK, describes how removal of crops from the ground represents a form of mineral nutrient mining and discusses the finite nature of the mineral sources of plant nutrients. In Chapter 9, by Paul Hallett and colleagues from the James Hutton Institute in Dundee, Scotland, and the University of Cottbus, Germany, the problem of restoring degraded soils to agricultural production is considered and potential solutions are presented.

Finally, we wish to acknowledge the help and advice provided by Colin Campbell and, particularly, by Paul Hallett, both of the James Hutton Institute, in the formative stages and throughout the commissioning of the chapters of this book. We believe the book will be found informative by and of interest to a wide range of users, from agronomists and those professionally engaged in soil science and agriculture, through to policy makers and to students in environmental sciences, food science and environmental and resource management courses.

Ronald E. Hester
Roy M. Harrison

Contents

Issues in Environmental Science and Technology No. 35
Soils and Food Security
Edited by R.E. Hester and R.M. Harrison

Published by the Royal Society of Chemistry, www.rsc.org

Editors

Ronald E. Hester, BSc, DSc (London), PhD (Cornell), FRSC, CChem

Ronald E. Hester is now Emeritus Professor of Chemistry in the University of York. He was for short periods a research fellow in Cambridge and an assistant professor at Cornell before being appointed to a lectureship in chemistry in York in 1965. He was a full professor in York from 1983 to 2001. His more than 300 publications are mainly in the area of vibrational spectroscopy, latterly focusing on time-resolved studies of photoreaction intermediates and on biomolecular systems in solution. He is active in environmental chemistry and is a founder member and former chairman of the Environment Group of the Royal Society of Chemistry and editor of 'Industry and the Environment in Perspective' (RSC, 1983) and 'Understanding Our Environment' (RSC, 1986). As a member of the Council of the UK Science and Engineering Research Council and several of its sub-committees, panels and boards, he has been heavily involved in national science policy and administration. He was, from 1991 to 1993, a member of the UK Department of the Environment Advisory Committee on Hazardous Substances and from 1995 to 2000 was a member of the Publications and Information Board of the Royal Society of Chemistry.

Roy M. Harrison, BSc, PhD, DSc (Birmingham), FRSC, CChem, FRMetS, Hon MFPH, Hon FFOM, Hon MCIEH

Roy M. Harrison is Queen Elizabeth II Birmingham Centenary Professor of Environmental Health in the University of Birmingham. He was previously Lecturer in Environmental Sciences at the University of Lancaster and Reader and Director of the Institute of Aerosol Science at the University of Essex. His more than 300 publications are mainly in the field of environmental chemistry, although his current work includes studies of human health impacts of atmospheric pollutants as well as research into the chemistry of pollution phenomena. He is a past Chairman of the Environment Group of the Royal Society of Chemistry for whom he has edited 'Pollution: Causes, Effects and Control' (RSC, 1983;

Fourth Edition, 2001) and 'Understanding our Environment: An Introduction to Environmental Chemistry and Pollution' (RSC, Third Edition, 1999). He has a close interest in scientific and policy aspects of air pollution, having been Chairman of the Department of Environment Quality of Urban Air Review Group and the DETR Atmospheric Particles Expert Group. He is currently a member of the DEFRA Air Quality Expert Group, the Department of Health Committee on the Medical Effects of Air Pollutants, and Committee on Toxicity.

List of Contributors

Soils and Food Security: Challenges and Opportunities

Peter Gregory, East Malling Research, New Road, East Malling, Kent, ME19 6BJ, UK, Email: Peter.Gregory@emr.ac.uk

Global Soils: Preserving the Capacity for Food Production

Luca Montanarella, European Commission, DG JRC, Via E. Fermi, 2749 I-21027 Ispra (VA), Italy, Email: luca.montanarella@jrc.ec.europa.eu

Soil Natural Capital and Ecosystem Service Delivery in a World of Global Soil Change

David A. Robinson (corresponding author), NERC Centre for Ecology and Hydrology, Environment Centre Wales, Deiniol Road, Bangor, LL57 2UW, UK, Email: darearthscience@yahoo.com
Bridgett A. Emmett, NERC Centre for Ecology and Hydrology, Environment Centre Wales, Deiniol Road, Bangor, LL57 2UW, UK
Brian Reynolds, NERC Centre for Ecology and Hydrology, Environment Centre Wales, Deiniol Road, Bangor, LL57 2UW, UK
Ed Rowe, NERC Centre for Ecology and Hydrology, Environment Centre Wales, Deiniol Road, Bangor, LL57 2UW, UK
Dave Spurgeon, NERC Centre for Ecology and Hydrology, Wallingford, Oxon, UK
Aidan M. Keith, NERC Centre for Ecology & Hydrology, Lancaster Environment Centre, Library Avenue, Bailrigg, Lancaster, UK
Inma Lebron, NERC Centre for Ecology and Hydrology, Environment Centre Wales, Deiniol Road, Bangor, LL57 2UW, UK
Neal Hockley, School of Environment, Natural Resources and Geography, Bangor University, Bangor, UK

Issues in Environmental Science and Technology No. 35
Soils and Food Security
Edited by R.E. Hester and R.M. Harrison

Published by the Royal Society of Chemistry, www.rsc.org

The Evaluation and Reporting of Soils in Sustainable Agriculture and Food Systems

Alfred E. Hartemink (corresponding author), University of Wisconsin, Department of Soil Science, F.D. Hole Soils Lab, 1525 Observatory Drive, Madison, WI 53706, USA, E-mail: hartemink@wisc.edu
Amanda L. Raster, University of Wisconsin, Department of Horticulture, 1575 Linden Drive, Madison, WI 53706, USA, E-mail alraster@wisc.edu
Molly M. Jahn, University of Wisconsin, Laboratory of Genetics and Department of Agronomy, 425 Henry Mall, Madison, WI 53706, USA, E-mail mjahn@wisc.edu

Agrobiodiversity and Potential Use for Enhancing Soil Health in Tropical Soils of Africa

Fredrick O. Ayuke (corresponding author), Department of Land Resource Management & Agricultural Technology, University of Nairobi, P.O Box 29053-00625, Nairobi, Kenya, Email: fredrick.ayuke@yahoo.com
Nancy K. Karanja, Department of Land Resource Management & Agricultural Technology, University of Nairobi, P.O Box 29053-00625, Nairobi, Kenya
Julius J. Okello, Department of Agricultural Economics, University of Nairobi, P.O Box 30197-00100, Nairobi, Kenya
Peter M. Wachira, School of Biological Sciences, University of Nairobi, P.O Box 30197-00100, Nairobi, Kenya
Gerald K. Mutua, Department of Plant Sciences, University of Nairobi, P.O Box 30197-00100, Nairobi, Kenya
David K. Lelei, World Agroforestry Centre, P.O Box 30677, Nairobi, Kenya
Charles K. K. Gachene, Department of Land Resource Management & Agricultural Technology, University of Nairobi, P.O Box 29053-00625, Nairobi, Kenya

Organic Matter Availability and Management in the Context of Integrated Soil Fertility Management in sub-Saharan Africa

Bernard Vanlauwe, International Institute of Tropical Agriculture, c/o ICIPE, PO Box 823 – 00621, Nairobi, Kenya, Email: b.vanlauwe@cgiar.org

Climate Change and Forest Dynamics: A Soils Perspective

Wendy Peterman (corresponding author), Conservation Biology Institute, 136 SW Washington Avenue, Suite 202, Corvallis, OR 97330, USA, Email: wendy@consbio.org
Dominique Bachelet, Conservation Biology Institute, 136 SW Washington Avenue, Suite 202, Corvallis, OR 97330, USA

Plant Nutrients

David Manning, School of Civil Engineering and Geosciences, Newcastle University, Newcastle upon Tyne, NE1 7RU, UK, Email: david.manning@newcastle.ac.uk

Soil Physical Degradation: Threats and Opportunities to Food Security

Paul D. Hallett (corresponding author), The James Hutton Institute, Invergowrie, Dundee DD2 5DA, Scotland UK, Email: paul.hallett@hutton.ac.uk

Kenneth W. Loades, The James Hutton Institute, Invergowrie, Dundee DD2 5DA, Scotland UK

Julia Krümmelbein, Lehrstuhl Geopedologie und Landschaftsentwicklung, Brandenburgische Technische Universität, Konrad-Wachsmann-Allee 6, 03046 Cottbus, Germany

Soils and Food Security: Challenges and Opportunities

PETER J. GREGORY

East Malling Research, New Road, East Malling, Kent, ME19 6BJ, UK, and Centre for Food Security, School of Agriculture, Policy and Development, University of Reading, Reading, RG6 6AR, UK
E-mail: peter.gregory@emr.ac.uk

ABSTRACT

Soils most obviously contribute to food security in their essential role in crop and fodder production, so affecting the local availability of particular foods. They also have a direct influence on the ability to distribute food, the nutritional value of some foods and, in some societies, the access to certain foods through local processes of allocation and preferences. The inherent fertility of some soils is greater than that of others, so that crop yields vary greatly under semi-natural conditions. Husbandry practices, including the use of manures and fertilisers, have evolved to improve biological, chemical and physical components of soil fertility and thereby increase crop production.

The challenge for the future is to sustain soil fertility in ways that increase the yield per unit area while simultaneously avoiding other detrimental environmental consequences. This will require increased effort to develop practices that use inputs such as nutrients, water and energy more efficiently. Opportunities to achieve this include adopting more effective ways to apply water and nutrients, adopting tillage

Issues in Environmental Science and Technology No. 35
Soils and Food Security
Edited by R.E. Hester and R.M. Harrison

Published by the Royal Society of Chemistry, www.rsc.org

practices that promote water infiltration and increase of organic matter, and breeding to improve the effectiveness of root systems in utilising soil-based resources.

1 The Role of Soils in Food Security

There are many definitions of food security, but one that is most commonly employed is that food security is the state when "all people, at all times, have physical and economic access to sufficient, safe, and nutritious food to meet their dietary needs and food preferences for an active and healthy life".[1] Food security is, then, a social construct in which availability, accessibility and utilisation all contribute to its achievement (Figure 1).[2,3] Food security is underpinned by effective food systems, which constitute a set of dynamic interactions between and within biogeophysical and human environments. Food systems comprise a number of activities (producing food; processing, packaging and distributing food; and retailing and consuming food) that lead to a number of associated outcomes, some of which contribute to food security (*i.e.* food availability, access to food and food utilisation), and others which relate to environmental and other social welfare concerns.[3] Because food security is diminished when food systems are disrupted or stressed, food security policy must address the whole food system.

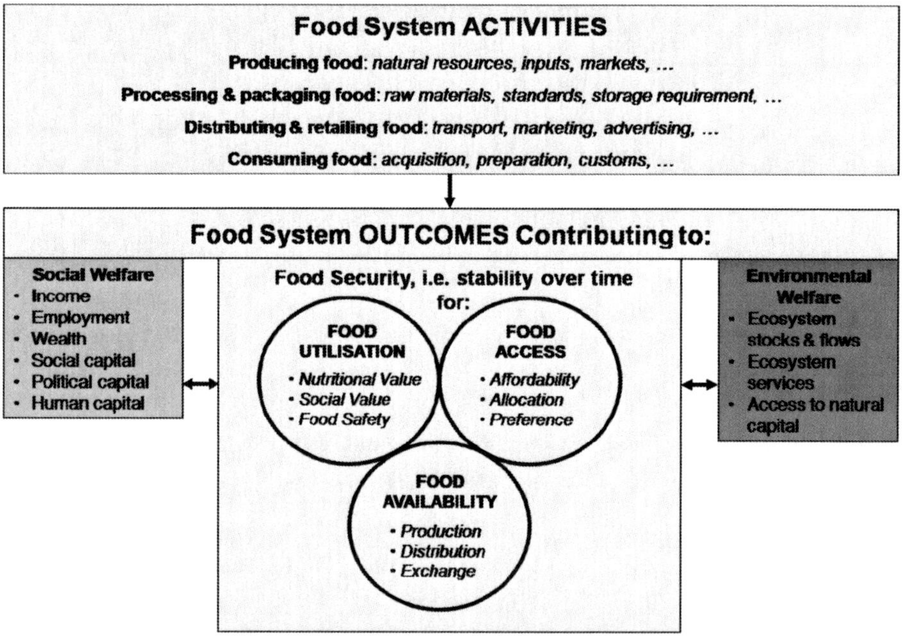

Figure 1 Elements of food security. (Source: adapted from Ericksen).[3]

Soils most obviously contribute to food security in their essential role in crop and fodder production, thereby markedly influencing the availability of food. The inherent properties of different soils have marked effects on crop productivity (see, for example, the writings of Cato and Pliny the Elder) and, while interventions to improve fertility can over-ride these properties, some soils are inherently more fertile and productive than others.[4,5] However, soils also have a direct influence on the ability to distribute food, the nutritional value of some foods and, in some societies, the access to certain foods through local processes of allocation and preferences. An obvious, if slightly extreme, example of the influence of soils on the ability to distribute food is seen in the behaviour of soils containing large amounts of swelling and shrinking clays (vertisols). These soils are frequently inherently fertile but are often very wet or waterlogged in one season making it impossible to harvest crops or to move easily across their surface, while in the dry season the shrinking of the soil induces large cracks so that engineered structures such as houses and irrigation ditches fail. The combination of shrinkage in the dry season followed by considerable swelling in the wet season means that roads are also difficult to sustain and the distribution of food can be affected.[6,7]

The nutritional value of many foods is markedly influenced by the soils on which they are grown, although processed foods are often supplemented with essential minerals and vitamins to make good any deficiencies. Crop production depends on the availability of sufficient quantities of the 14 essential mineral elements required for plant growth and reproduction.[8] These essential nutrients include the macronutrients required in large amounts by plants (nitrogen (N), phosphorus (P), potassium (K), calcium (Ca), magnesium (Mg) and sulfur (S)) and the micronutrients (boron (B), chlorine (Cl), copper (Cu), iron (Fe), manganese (Mn), molybdenum (Mo), nickel (Ni), and zinc (Zn)) which are required in smaller amounts.[8,9] Deficiency in any one of these elements restricts plant growth and reduces crop yields, so that they are often applied to crops as inorganic or organic fertilisers to increase crop production [9]. Humans require many more mineral elements for their wellbeing than plants.[10–12] In addition to the 14 elements essential for plants, humans also require significant amounts of cobalt (Co), iodine (I), selenium (Se) and sodium (Na) in their diet and, possibly, small amounts of arsenic (As), chromium (Cr), fluorine (F), lead (Pb), lithium (Li), silicon (Si) and vanadium (V). The majority of these mineral elements are supplied to humans by plants.

Unfortunately, the diets of over two-thirds of the world's population lack one or more of these essential mineral elements,[13,14] with over 60% being Fe-deficient, over 30% Zn-deficient, almost 30% I-deficient, and about 15% Se-deficient. Dietary deficiencies of Ca, Cu and Mg are also prevalent in many countries. This mineral malnutrition is attributable to either crop production on soils with low phytoavailability of the mineral elements essential to human nutrition, or consumption of staple crops, such as cereals, or phloem-fed tissues, such as fruit, seeds and tubers, that have inherently low tissue concentrations of certain mineral elements, or both.[15,16] Soils that are low in phytoavailable minerals include:[10]

- alkaline and calcareous soils that have low availabilities of Fe, Zn and Cu; these comprise 25–30% of all agricultural land;[17–19]
- coarse-textured, calcareous or strongly acidic soils that have low Mg content;[20]
- mid-continental regions that have low I content;[21,22] and
- soils derived mostly from igneous rocks that have low Se content.[23,24]

In contrast, excessive concentrations of potentially toxic mineral elements may also compromise both crop production and human health.[25] On acid soils occupying about 40% of the world's agricultural land, toxicities of Mn and aluminium (Al) may limit crop production, while on sodic or saline soils (5–15% of agricultural land) sodium (Na), B and Cl toxicities frequently reduce crop production, and toxicities of Mn and Fe can occur in waterlogged or flooded soils.[17,25] Excessive concentrations of Ni, Co, Cr and Se can limit growth of plants on soils derived from specific geological formations (*e.g.* serpentine)[17] and toxic concentrations of As, cadmium (Cd), Cu, mercury (Hg), Pb and Zn have accumulated in agricultural soils in some areas due to mining and industrial activities.[17] These toxic elements, contained in plants and animals that graze on them, can accumulate in the food chain with detrimental consequences for human health.

Soils also directly influence elements of accessibility and social preferences for certain foods. An obvious European example is the importance of "terroir" in the perceived quality of certain wines and the social cachet attached to them.[26] In a multi-factorial experiment, it was demonstrated that the effect of soil appeared to be principally *via* effects on vine water status rather than effects on mineral nutrition.[26] Similarly, in Asia, different rice types associated with different soils and growing systems have assumed positions of political, social and commercial importance.[27]

2 Key Soil Constraints to Crop and Fodder Production

The concept of soil fertility is widely used as a framework for exploring the relationships between crop productivity and soil characteristics. It is an expression that synthesises chemical, physical and biological properties of soils and their effects on the growth and activities of root systems and the shoot.[28] Soils can be inherently fertile because of combinations of high mineral nutrient availability, good soil structure, high available water contents and appropriate microbial and faunal communities that facilitate good root and shoot growth, or be managed to promote soil fertility through, for example, cultivation techniques that do not destroy structure or through additions of manures and fertilisers.[29] More recently, crop genotypes have been developed to overcome some key soil constraints to fertility.

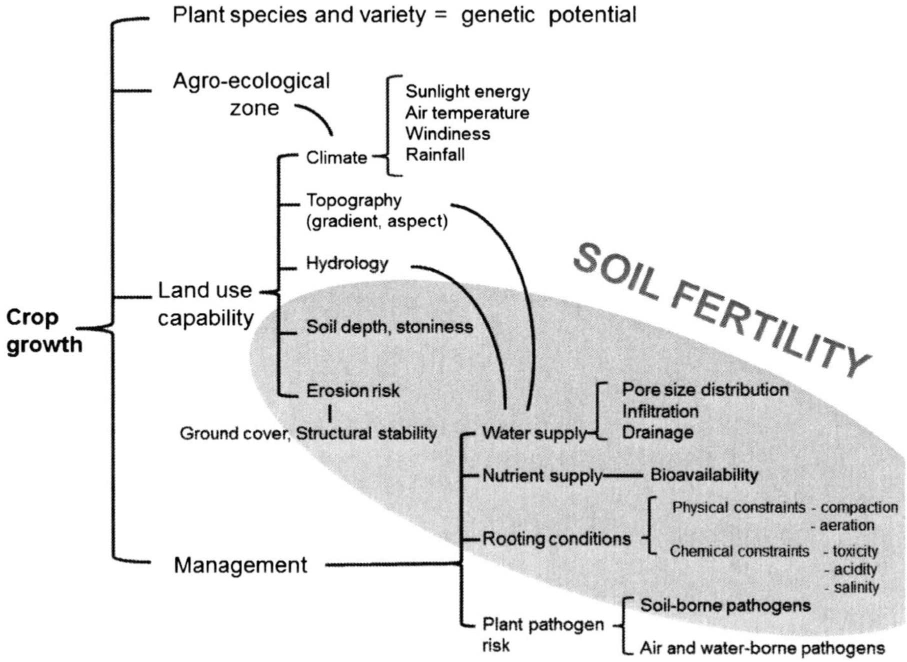

Figure 2 Figurative summary of components of soil fertility. (Source: adapted from Stockdale).[29]

In essence, the task of farmers and their advisers is to identify the soil constraints to crop production, and then to ameliorate these with inputs and/or management practices that minimise them so that the potential yield determined by genetic and climatic properties can be approached (Figure 2). The main factors limiting yields on many soils are: depth of soil, soil compaction, water supply, nutrient supply, erosivity, and soil reaction, including pH and salinity. These key soil constraints vary between soils and their past histories of use and management. For example, salinity is often a constraint to production on soils irrigated with low quality water and with no or limited drainage, whereas human-induced compaction may constrain production on soils cultivated with heavy machinery.

2.1 Soil pH

Many aspects of soil chemistry, and hence soil fertility, are influenced by soil pH, including the bioavailability of plant mineral nutrients, microbial activity and root growth.[25] Nearly all natural soils have a pH between 4 and 10. Soil pH at a given location is a function of soil composition (the relative proportions and types of organic and mineral constituents) and the consequent ion-exchange and hydrolysis reactions. Generally, soil pH values of <4 are uncommon because in such acid soils aluminosilicate and oxide minerals dissolve and buffer the pH.

However, in soils recently reclaimed from marine sediments, the presence of sulfur in appreciable quantities can give rise to very acidic soils on drainage (acid sulfate soils) that severely curtail plant growth.[30] Slight-to-moderately alkaline soils are typically associated with calcareous parent materials or accumulations of calcium carbonate, although some may contain magnesium carbonate. Highly alkaline soils (pH 8.5 to 10) are usually associated with the presence of dissolved sodium carbonate, which results in the greater production of OH^- ions than calcium carbonate. Such soils are typically associated with irrigated soils in arid regions.

The optimum soil pH for most crops is typically 6.5 to 7, a pH at which the availability of most plant mineral nutrients is optimal.[8] Because rain is naturally slightly acidic, and many industrial processes result in acid deposition on vegetation and land surfaces, soil acidification is the norm. In many regions, soil pH is frequently managed by the application of materials such as lime and dolomite that act to neutralise the acidity. The growth of a wide range of crop and pasture species is enhanced by this practice, with responses ranging from 5% to 200% relative to an unlimed control.[31] This wide range of response is due to a combination of soil and crop factors which mean that there is no unique relation between observed response to lime applications and meaningful soil parameters.[31]

Naturally occurring acid soils with pH <5.5 in the upper soil layers occupy about 30% (4000 million ha) of the world's ice-free land area. They are found predominantly in two geographical regions: one in the humid northern temperate zone and the other (the majority) in the humid tropics;[25] these are areas where high rainfall leads to intensive leaching of basic cations such as calcium and magnesium, so that less soluble and acidic minerals, such as aluminium and iron oxides, become more dominant. The largest areas of acidic soils are located in North and South America (41% of the world's soils by area), Asia (26%) and Africa (17%). Because of the difficulties of growing crops on such soils without inputs of liming materials and phosphatic fertilisers (aluminium and iron oxides absorb phosphate, strongly limiting its bioavailability), many of the world's acidic soils are still under natural vegetation. A notable exception to this has been the recent development of the Cerrado region of Brazil in which amelioration of subsoil acidity with lime and the use of phosphate fertilisers has allowed productive agriculture to occur.[32]

In addition to natural processes of soil acidification, some agricultural practices can also enhance the process. In particular, the use of ammoniacal fertilisers is well known as a factor leading to acidification because the nitrification of ammonium by microorganisms results in the production of protons.[33] Similarly, the process of nitrogen fixation by free-living bacteria and bacteria living in the nodules of leguminous crops also results in acidification through the production of protons.[34]

2.2 Saline and Sodic Soils

About 7% of the world's total land area (930 million hectares) is salt-affected.[35] Salt accumulation, as a consequence of irrigation accompanied by limited or no drainage, has been associated with the decline of several past civilisations, including those of Babylon, Carthage and the Hohokam Indians.[36] Salts, particularly of sodium, magnesium and calcium, accumulate in soils because plant roots are selective in the ions that they allow into the plant, and these ions are required in smaller quantities than many others. In many circumstances this is not a problem if there is sufficient rainfall (or irrigation) to leach the accumulated salts from the soil, but in arid regions, or in areas where irrigation with water-containing dissolved salts is practised, salts build up.

Saline soils are those for which the electrical conductivity of a saturated paste extract exceeds 4 dS m^{-1}, corresponding to an osmotic potential of about -145 kPa or a total cation concentration of about 40 mmol l^{-1}.[25] This approximates to values at which the growth of many plants is reduced by salt accumulation. In practice, measuring electrical conductivity of saturated paste extracts is laborious, so a soil–water (1 : 5) extract is normally used and a conversion factor applied. In contrast to saline soils, there is no universally applied definition of what constitutes a sodic soil, although the key factor is a high proportion of sodium relative to other cations. Sodicity is typically defined in relation to the exchangeable sodium percentage, which expresses the sodium on exchange sites as a percentage of the total exchangeable cations. In the USA, a value of 15% is typically used to define a sodic soil, but in Australia a value of 6% is common. The difference arises because the property of agricultural importance most frequently associated with sodicity, the dispersion of soil, varies considerably with soil type and is not uniquely related to exchangeable sodium percentage.

Naturally occurring saline soils are widespread in arid and semi-arid regions where low rainfall and no, or limited, leaching occurs to remove salts from soils (in many ways, these are the opposite conditions from those that lead to the formation of the acidic soils described in section 2.1). In many Middle Eastern countries, saline soils have formed as a consequence of irrigation with water over many hundreds of years, but in Australia dryland salinity is a more complex phenomenon.[37] Salt has accumulated deep in soils of many parts of Australia for prolonged periods, but this was not a problem until large-scale land clearance and deforestation replaced deep-rooted perennial species with shallow-rooted annual crops, allowing increased drainage of water to the watertable. This led, in turn, to rising watertables and the upward movement of salts through the soil profile into the rooting zone of the annual crops and pastures, with severe consequences for productivity.[38]

Irrigation-induced salinity is increasing along with the introduction of new irrigation schemes. Most commonly, it arises because of inattention to the need for adequate drainage systems without which the consequences are combinations of rising watertables, upward movement of salts from naturally-occurring saline soil layers, and accumulation of toxic concentrations of salts. Even when drainage is adequate, inadequate leaching can result in the build-up of high salt concentrations, with deleterious effects on plant growth. With pressure to use effluents and "grey water" for irrigation increasing around the world, many of which contain relatively high concentrations of salts, including sodium, there is an increased risk of salinisation of soils.

The response of plant growth to salts differs markedly between species, ranging from highly sensitive species such as rice, through moderately tolerant species such as barley and alfalfa, to highly tolerant (halophytic) species such as saltbush (*Atriplex amnicola*).[39] Often saline and sodic soils affect plant growth through one or more sets of processes: (i) physiological drought, resulting from a reduction in water-availability caused by reduced osmotic potential; (ii) ionic imbalance in plant cells; and (iii) toxicity due to specific ions, such as sodium, chloride or borates. In addition, high pH may also result in plant stress and have effects *via* nutrient availability and plant metabolism. In controlled experiments, plant-growth responses to salt stress are frequently observed as a two-phase process.[40] Initially, the response can be attributed to osmotic stress but as the exposure to salt increases, salt-specific effects in the older leaves of plants result in premature leaf senescence and associated reductions in leaf photosynthetic area, leading to reduced plant growth. The initial response to salt is due to osmotic stress which causes cell dehydration and shrinkage in the root, reducing the ability of the plant to take up water; if the salt is removed, this is a transient effect that can be reversed, but under typical field conditions the plant must expend increasing amounts of energy to maintain turgor. For example, in a sandy loam, plants under non-saline conditions were able to maintain turgor with a water content of only 5%, whereas at an electrical conductivity in a soil–water (1 : 5) extract of 1 dS m^{-1}, a water content of at least 18% was required to prevent plants wilting.[25]

Crop plants exhibit a variety of mechanisms that confer some degree of tolerance to saline or sodic conditions, including: (i) selective exclusion/inclusion of ions; (ii) induction and up-regulating of antioxidative enzymes; (iii) accumulation of certain organic solutes to increase hyperosmotic tolerance; and (iv) efficient salt excretion *via* salt glands or tissue shedding. Typically, salt-tolerant plants have slower rates of sodium and chloride ion transport to leaves than sensitive plants and a greater ability to compartmentalise these ions into vacuoles, so preventing their accumulation in the cytoplasm or cells walls.[39] Tolerance of salt is not a simple property and is a genetically complex trait affected by many genes with additive, dominant and reciprocal effects.[41,42]

2.3 Soil Strength and Structure

The rate of root growth, and hence the ability of plants to exploit water and nutrient reserves in soils, is affected considerably by soil strength.[43] This strength is a consequence of capillary forces and bonds between particles in soils. Management practices resulting in compaction or loss of organic matter generally increase soil strength and decrease root growth. The arrangement of soil materials, gases and water determines soil structure, which has been defined as the "spatial heterogeneity of the different components or properties of soil".[44] Soil structure includes the complex interactions between the many minerals, organic substances, organisms, gases and water in soils at a range of organisational levels, ranging from the interactions between clay particles and domains (typically 10^{-7} m in diameter) to large-scale cracks and field-scale hydrological processes (often >100 m in size). Within this physical framework, multiple chemical and biological processes affecting soil fertility occur and roots function to support the growth of the whole plant.

Soil structure is created by a range of biological, hydrological and mechanical processes, often intimately associated with the growth and activity of root systems. Structure is formed over time by a combination of dispersion and aggregation of soil particles, driven by cycles of wetting and drying, additions of organic compounds, faunal burrowing, root penetration and the bonding of soil mineral surfaces. These processes are accentuated at the root-soil interface, resulting in a more distinct and physically stable structure than in the bulk soil. Roots mechanically deform soils as they pass through them and the flora, fauna and exudates associated with them alter bond energies between particles, water surface tension and soil physical behaviour.[45] Structure formation is evident from the sheath of soil (sometimes referred to as a "rhizosheath") that generally adheres to excavated roots and has been shown in many studies to be more structurally stable than bulk soil.[46,47] Root hairs also play an important role in bonding soil to root surfaces, which increases contact and hence the potential uptake of water and nutrients.[48,49] Secondary metabolites from soil microbes are also a major driver of structure formation close to roots, with arbuscular mycorrhizal fungi also playing a role.[50,51]

Alternate cycles of water extraction by roots and wetting by rain induce shrinking and swelling in many soils that can cause them to crack and slake.[52] Such cracks form large pores (macropores) in soil that act as pathways for rapid transmission of gases and water between upper and lower soil layers; in this way, water and dissolved solutes can be moved rapidly downwards to groundwater, by-passing the bulk of the soil. However, plant roots can also exploit large pores as means of by-passing obstructions, such as compacted soil layers induced by inappropriate cultivation (*e.g.* plough pans), thereby gaining access to nutrients and water in subsoils.[53]

The ability of roots to deform soil during elongation influences the mechanical resistance to root penetration and, ultimately, the size of the root system and the performance of crops and pastures. Several studies have measured the distribution of soil porosity or particles after a root has penetrated a soil core and showed changes in bulk density at up to 4–5 mm away from the root surface.[54] For maize roots growing in sand, particle image velocimetry has shown density increases of up to 30% adjacent to growing root tips, with an approximately exponential variation in particle displacement as a function of distance from the root surface.[55] Local variation in sand density was associated with root cap frictional properties, so that roots with a root cap shedding mucilage and border cells deformed the soil radially, with density increases generally confined to the flanks of the root, whereas roots from a capless mutant that shed neither mucilage nor border cells had zones of greater density in front of the root tip.

The degree of particle displacement around roots is important not only for determining soil physical properties but also the degree of root-soil contact and hence the accessibility of nutrients and water. In water-saturated and heavily compacted soils, problems with gas exchange can occur, while incomplete root-soil contact due to soil structure or root shrinkage can reduce the uptake of water and nutrients such as nitrate.[56] Root-soil contact has been known for a long time to affect the growth rates of crops, with the production of a good tilth being a prime requirement of seedbed formation. New techniques, such as X-ray computed tomography, are increasingly being used as a non-invasive methodology to visualise and quantify these interactions.[57]

3 Contributions to Recent Increases in Crop Production

As discussed in section 1, while food production alone cannot guarantee food security, it is an essential component of food systems. During the last 60 years, the world's human population has increased from about 2.2 billion in 1950 to about 7 billion in 2011 and this has been sustained by substantial increases in crop and animal production.[58] Only about 3 billion of the world's 13.4 billion hectare land surface is suitable for crop production and about one half of this is already cultivated (1.4 billion ha in 2008).[59] It is widely recognised that, globally, only a small proportion of future increases in crop production will come from the cultivation of new land (about 20%), with the majority coming from intensification *via* increased yield (67%) and higher cropping intensity (12%; Table 1).[60,61] This means that *per capita* arable land area will continue to decrease (it decreased from 0.415 ha in 1961 to 0.214 ha in 2007) while average cereal yield will need to increase by about 25% from 3.23 t ha^{-1} in 2005–07 to 4.34 t ha^{-1} in 2030.[59,61]

Yield results from the interaction of three factors: genotype (G) × environment (E) × management (M).[62] Evans highlights how the synergistic effects of these interactions, linked to innovative technologies, have

Table 1 Projected contributions (%) to increased crop production between 1997–99 and 2030. (Source: adapted from Bruinsma).[61]

	Land area expansion	*Increase in cropping intensity*	*Yield increase*
All developing countries	21	12	67
Sub-Saharan Africa	27	12	61
Near East/North Africa	13	19	68
Latin America and Caribbean	33	21	46
South Asia	6	13	81
East Asia	5	14	81

contributed to past increases in yield.[63,64] Among the important contributors to these have been:

- improved germplasm, able to grow vigorously (*e.g.* hybrids), resist pathogens, and respond to fertilisers without lodging (in particular the use of dwarfing and semi-dwarfing genes in rice and wheat);
- the application of fertilisers and, particularly, the availability of affordable nitrogen fertiliser;
- the development of chemicals to control weeds, pests and diseases; and
- improved irrigation systems, especially in rice-producing countries and for some previously rain-fed crops.

These technological innovations, together with institutional and market reforms, have modified G, E and M to greatly increase yields. As a result of these innovations, over large areas the yields of many crops have increased year-on-year (Figure 3). Global yield increases for a number of crops have typically been linear (Figure 3a), with values of 53 kg ha^{-1} yr^{-1} for rice, 41 kg ha^{-1} yr^{-1} for wheat, and 63 kg ha^{-1} yr^{-1} for maize over the period 1961 to 2004. Increases in yield have also been linear with time in many individual countries (*e.g.* wheat yields in several European countries)[65] although, in a few instances, technological innovations have produced more rapid, step-wise increases in yield (*e.g.* Australia; Figure 3b).[66] Although crop yields have increased globally and throughout North and South America, Europe, Australia and much of Asia, a notable exception has been that of Africa where, for example, *per capita* food production decreased by about 5–10% between 1980 and 1995.[67] The reasons for the poor performance in Africa relative to other countries are many and include social unrest and war, poor institutions and governance, climatic variability making reliable irrigation difficult, and weathered soils that are deficient in nutrients. It has been argued by many that a major factor behind many of the observed decreases in yield in Africa countries has been the decline of soil fertility accompanied by the lack of fertiliser application. Much has been written about the need to "re-capitalise" the soils of Africa, especially with regard to P status, but progress has been limited.[67,68]

Because the area of cropped land is likely to increase proportionately less than the future demand for food, reducing the gap between current yields and potential yields is a major goal for the future.[69] Potential yield is a theoretical upper limit to yield imposed by solar radiation (affecting growth), temperature (affecting development and growth) and water supply (affecting mainly growth but also development). A review of data from crops of maize, rice and wheat grown in a range of countries showed that the gap between potential and the actual yields ranged from about 20 to 80%.[70] In many irrigated cereal systems, yield appeared to plateau at or about 80% of potential yield while in rain-fed systems, average yields were commonly 50% or less of potential.[70] While part of the yield gap is inevitable because of crop losses during harvest, storage and transport, and the way that land areas are reported,[69] there are still large

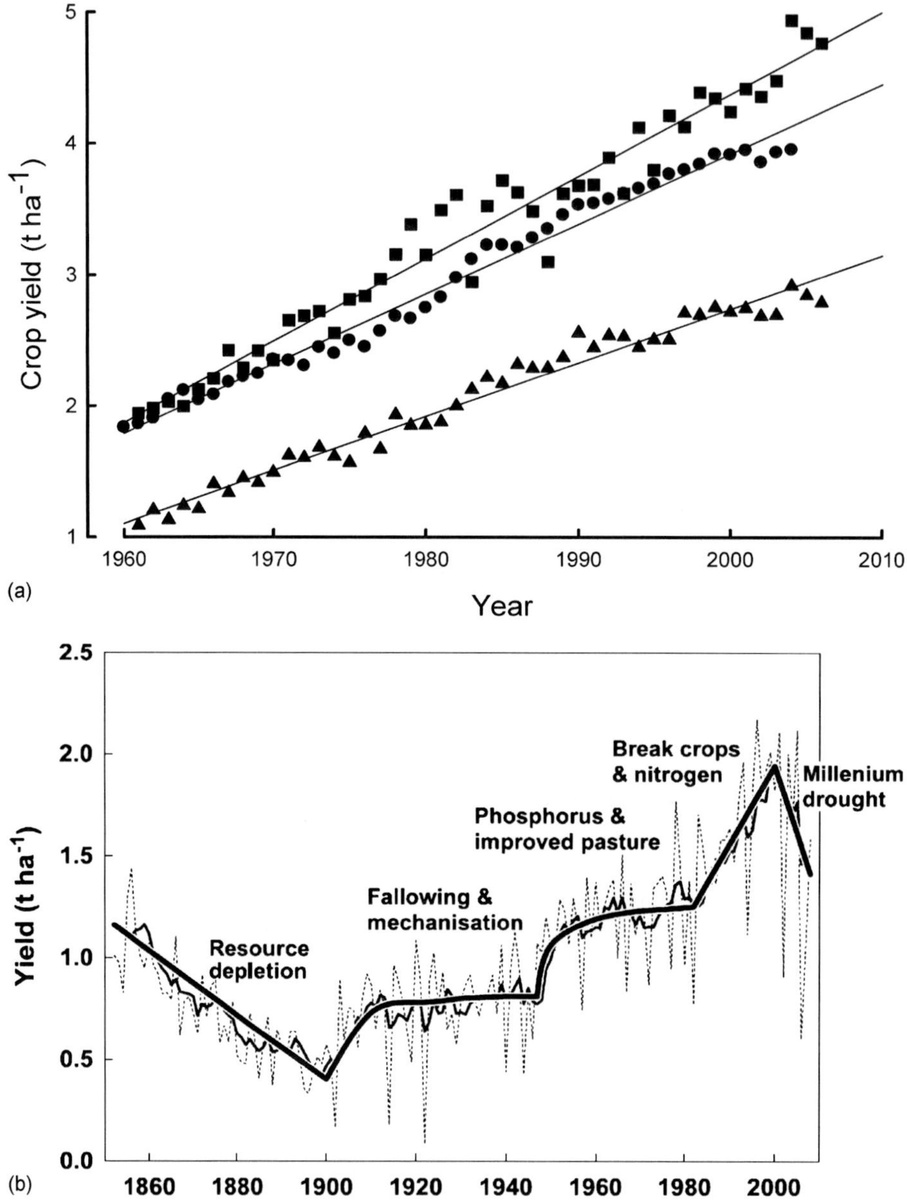

Figure 3 Changes in yields with time. (a) Global average annual yields (data from USDA) for maize (squares), rice (circles), and wheat (triangles); the lines are linear regressions. (b) National average annual yield of wheat in Australia (Source: modified from Angus);[66] the dashed line joins annual values, the thin solid line is a 5-year running average, and the black line summarises major trends in yield with respect to changes in crop husbandry.

differences in performance between adjoining farms. A fundamental constraint in many irrigated systems is the uncertainty in growing season weather; this is also a factor in rain-fed systems where interactions between water and nutrient availability are complex.[70,71] Raising yields above 80% of yield potential is possible, but only if technologies can be developed and adopted that reduce the uncertainties faced by farmers in assessing soil and climatic conditions, or respond dynamically to these conditions, or both (for example, installation of nutrient and water sensors).[70] Such technologies may have the added benefits of increasing the efficiency of use of inputs and reducing losses off-site as well as increasing yields.

Analysis of global wheat, maize and rice production found that yield gaps were significantly correlated with irrigation, market accessibility and influence, availability of agricultural labour, and slope; the contribution of these factors varied substantially between regions and generalisations as to the best means of reducing yield gaps were not possible.[72] In China, the potential to reduce yield gaps of maize was demonstrated on 66 on-farm experimental plots, raising yields to 13 t ha^{-1} on average (nearly twice the typical farmer yield) without any increase in N fertiliser application.[73] The demonstration of this potential was achieved using simulation models to identify appropriate combinations of planting date, crop density and cultivar at each site, based on long-term weather data, and then changing variety, sowing date and spacing as appropriate. This was combined with an in-season management strategy for nitrogen fertiliser that resulted in a greater proportion of the currently applied fertiliser being applied later in the growing season. This integrated agronomic approach, combined with cultivars of appropriate duration, increased yields and reduced off-site nutrient movement.[73] In addition to technological changes, other social and educational factors also contribute to the ability to increase yields. For example, an analysis of yield-gaps for rice in four intensively cropped regions in Indonesia, The Philippines, Thailand and Vietnam demonstrated that the farmers with the best yields were typically more educated and used fertilisers and labour more efficiently than others.[74] The importance of narrowing the yield gap between the average and the best farmers was highlighted by the conclusion that were this to be achieved then the resultant production would meet the projected increased demand by 2050 (assuming no change in diet) in all countries except the Philippines where other institutional changes would also be required.[74]

4 Opportunities for Sustainable Increases of Yield

The intensification of crop and animal production systems to meet human demands for food has often been achieved at some cost to other ecological goods and services. For example, excessive nutrient inputs, especially of nitrogen and phosphorus, have resulted in coastal eutrophication and reduced the quality of water in reservoirs used for drinking water.[75] Similarly, cultivating soils for crop production has often increased the frequency of substantial soil erosion by either water, tillage or wind so that the current mobilisation of soil globally is 35 ($\pm$10) Pg yr^{-1} or about 5 t yr^{-1} for every person on the planet.[76] In addition, the clearance of forests for agriculture has led to decreased biodiversity and substantial inputs of greenhouse gases to the atmosphere. This has led many to challenge the technologies that have resulted in today's intensified agriculture and to call for the development of sustainable production practices that will ensure that the multiple functions of land, and the many ecosystem goods and services provided by land, are conserved and sustained for future generations.[77–80]

Although there is general agreement that agricultural sustainability includes elements of profitable production, environmental stewardship and social responsibility, there is much less agreement as to how sustainability is to be achieved in practice beyond the need to integrate biological and ecological insights into the production process. A review of the literature suggested that sustainability encompassed four key principles that[78] "(i) integrate biological and ecological processes such as nutrient cycling, nitrogen fixation, soil regeneration, allelopathy, competition, predation and parasitism into food production processes; (ii) minimise the use of those non-renewable inputs that cause harm to the environment or to the health of farmers and consumers; (iii) make productive use of the knowledge and skills of farmers, thus improving their self-reliance and substituting human capital for costly external inputs; and (iv) make productive use of people's collective capacities to work together to solve common agricultural and natural resource problems, such as for pest, watershed, irrigation, forest and credit management". Such principles go well beyond the need for continued technological innovations, such as new germplasm underpinning increases in yield, and embrace the need to develop important capital assets for agricultural systems, including natural, social, human, physical and financial capital. A corollary of this analysis is that many disciplines and ways of thinking will be required to develop sustainable systems and that there is unlikely to be a single solution appropriate to all soils and production systems.

Soil is a component of the natural capital used by humans for food production, and the functions provided by soils are major contributors to almost all of the provisioning, regulating and cultural services provided by ecosystems.[77] But using soils for agriculture almost invariably leads to changes in soil properties such as nutrient status, pH, organic matter content and some physical properties as interventions are made to influence soil fertility. These changes, while beneficial to crop production, are often detrimental to other ecosystem services so that there is tension between the different functions undertaken by soils.[80] There is little doubt that, for most soils, sustainable production is inextricably linked to the maintenance of soil organic matter contents through appropriate additions to offset the losses caused by cultivation and nutrient depletion.[68,81] Soil organic matter influences many key processes including release of greenhouse gases (GHGs), nutrient cycling, microbial and faunal diversity, and many soil physical properties.[80] Several analyses have reached similar conclusions that in order to avoid severe degradation of the natural capital offered by soils, and to reduce GHG emissions, future systems for crop production will need to produce higher yields on existing cropland, limit expansion of the cultivated area, achieve a substantial increase in N fertiliser efficiency, and improve soil quality through increasing soil organic matter.[80,82]

An influential and wide-ranging assessment of future food and farming concluded that sustainable intensification was required to meet future demands for food, and that all technological means of achieving this should be assessed and appropriately utilised.[79] How best to achieve this intensification, and whether intensification will lead to land being spared for other ecological services, is a subject of considerable current debate and research. The extent to which past increases of crop yields have spared land for nature conservation is a matter of considerable debate because: (i) the on-farm losses of biodiversity due to practices giving high yields may outweigh the benefits of sparing biodiverse habitats; (ii) high-yielding crops may have negative effects on off-site biodiversity; and (iii) land-sparing does not occur or is imperfect. The complexity of the factors involved is indicated in an analysis of the changes in yields of 23 staple crops for 124 countries between 1979 and 1999.[83] While the *per capita* area of the 23 staple crops decreased in developing countries where large yield increases occurred, this was countered by a tendency for an increased area of non-staple crops, leading to only a weak tendency for land-sparing overall. In developed countries there was no evidence that higher yields reduced *per capita* cropped areas, probably because agricultural subsidies promoted production, thereby overriding any land-sparing effects. The study concluded that land-sparing is a weak process, but that improved agricultural technology may have contributed to the maintenance of natural vegetation cover in the past and that future conservation benefits, while debatable, are potentially available if land-use policies are also modified.[83]

The potential of different routes to achieving intensification to meet future food demands is a topic of on-going research, but several studies have indicated the importance of approaches that combine multiple disciplines that take account of local soil, ecological and societal conditions.[84,85] For example, alternative pathways to increasing yields to the required levels were investigated in one global study involving technological and educational advances.[86] The options examined were: (i) current technology, in which each economic group retained its present relation between yield and N fertiliser use; (ii) technological improvement, in which technological advances continue along existing temporal trends to 2050; (iii) technology transfer, in which low-yielding countries adopt and adapt the existing high-yielding technologies of high-yielding countries; and (iv) technology improvement and transfer, in which all countries achieve soil- and climate- adjusted yields. If present trends of intensification in rich nations and extensification in poor nations persist, then by 2050 an additional 1 billion ha of land would be cleared and greenhouse gas emissions would increase to 3 Gt yr^{-1} and N use to 250 Mt yr^{-1}. However, if intensification were concentrated on existing cropland, and transfer and adoption of high-yielding technologies were successful, then only 0.2 billion ha would be cleared, GHG emissions would be reduced to one-third (1 Gt yr^{-1}) and global N use would be 225 Mt yr^{-1}. Although this analysis omits any effects of future climate change, it indicates what might be possible with investment in innovative technologies, education and infrastructure.

4.1 Improved Efficiency of Resource Use

One aphorism that summarises the current thinking about future intensification of production is that we shall need to produce "more with less" and, in the case of irrigation, "more crop per drop". A major requirement is to produce higher yields with inputs that do not lead to environmental problems either on- or off-site.

4.1.1 Nutrients

Nutrient additions that are inadequate relative to crop offtake degrade land through nutrient mining, while additions that are excessive degrade land, water and air through leaching, eutrophication and gaseous emissions.[75] Ideally nutrient additions (whether as mineral fertilisers or manures) and soil biota should be managed to deliver nutrients to crops synchronously with demand,[87] but this has proved difficult to achieve in practice because applications must normally be made before the demand exists and large crop canopies do not permit application of solid sources to soils.

In developed countries, fertilisers are often applied in response to soil test results of available nutrients such as P and K, and with regard to the likely level of offtake for nutrients such as N.[29] Increasingly, models of crop and soil-nutrient dynamics are being employed in decision support systems to adjust applications to local conditions.[29] Furthermore, the rapid evolution of "precision agriculture" techniques in the last decade has allowed the application of chemical inputs at the sub-field scale.[88,89] Harvesters with weighing facilities and global positioning systems have allowed the production of yield maps on an almost routine basis and these, together with grid and transect sampling of soils to produce maps of nutrients, are allowing farmers to contemplate site-specific nutrient management.[90] Such management is knowledge-intensive and requires multiple forms of knowledge to be integrated in a way that can be practically managed. This is all a far cry from the situation in many developing countries where fertilisers are scarce and, when present, are often applied to poor effect.

In several regions, broader integrated approaches to the maintenance of soil fertility have proved capable of sustaining production for prolonged periods.[29] Such approaches involve inputs of organic materials in the form of crop residues, targeted use of legumes in the crop rotation, agro-forestry systems, animal manures, green manures, and dual-purpose legumes. Responses to additions of organic materials to soils are complex because such inputs can change the biomass, activity and diversity of soil organisms, leading to altered rates of decomposition and nutrient cycling, and to changes in soil structure and other physical properties.

The benefits of integrating both organic and mineral fertiliser applications (integrated nutrient management) have been demonstrated in many studies. For example, in an 8-year study of maize/millet and wheat/rice rotations in the mid-hills of Nepal,[91] farmyard manure and fertilisers were applied to the maize and wheat crops every year with the succeeding crops (finger millet and rice) utilising residual nutrients. Yields of maize, millet and rice were greater when manure rather than fertiliser was applied but yields of wheat were less. The combined application of manure and fertiliser significantly increased yields of maize and wheat compared with applications of either manure or fertiliser alone. Averaged over years, maize yielded significantly more (+563 kg ha^{-1}) when manure and fertiliser were applied than when manure alone was applied, and maize given manure yielded significantly more (+307 kg ha^{-1}) than maize receiving fertiliser alone (Table 2). Similarly, yields of millet were increased significantly by 630 and 705 kg ha^{-1} when either manure and fertiliser or manure alone, respectively, were applied to maize rather than inorganic fertiliser. Similarly, wheat yielded significantly more (+1183 kg ha^{-1}) when manure and fertiliser were applied than when manure alone was applied but, in contrast to maize, wheat given manure yielded less (−594 kg ha^{-1}) than that receiving fertiliser alone. The difference between maize and wheat in their yield responses to manure and fertiliser may reflect the different timings of applications; fertilisers may provide a more available source of nutrients to

Table 2 Average differences in grain yield (kg ha^{-1}) and their standard error for particular contrasts between treatments applied over eight years to maize grown in a maize/millet rotation in Nepal. (Source: adapted from Sherchan).[91]

	Maize		*Millet*	
Treatment comparisons†	*Yield difference*	*Stan. error*	*Yield difference*	*Stan. error*
Manure + fertiliser *vs.* manure	563***	72.0	−73**	23.6
Manure + fertiliser *vs.* fertiliser	869***	127.5	631***	78.0
Manure *vs.* fertiliser	307*	142.2	705***	87.2
Nitrogen top-dressing *vs.* no nitrogen top-dressing	797***	217.5	68***	18.8
No lime *vs.* lime	107*	47.7	135 ns	103.6
High rate of manure *vs.* low rate of manure	418**	145.5	107 ns	104.9
No basal fertiliser *vs.* basal fertiliser	304 ns	194.2	166 ns	385.3

†For descriptions of manure and fertiliser treatments see ref. 91; *, ** and *** indicate significance at the 5%, 1% and 0.1% levels, respectively; ns = non-significant.

crops such as wheat grown in the cold dry winter than to crops such as maize grown in the wet summer, whereas organic materials may provide a more available source of nutrients in the wet season when decomposition is more rapid.[91] For the rice and millet crops grown on residual nutrients, integrated nutrient treatments yielded more than fertiliser applications alone, and slightly less than with manure treatments alone. On average, treatments integrating both organic and mineral fertilisers yielded 35% more for both crops in the maize/millet rotation and 16% more for both crops in the wheat/rice rotation compared with the treatments where inorganic fertiliser alone had been applied.

Despite the demonstrable benefits of integrated nutrient management from a scientific viewpoint, there are, however, several limitations to its adoption in practice. Two important limitations are the availability of suitably nutrient-rich sources of organic materials and the availability of labour to both generate the organic materials and to spread them. In the mid-hills of Nepal, tree leaves and crop straw are fed to cattle to generate farmyard manure,[92] but partial cost-benefit analysis favours fertiliser application because of the large labour costs involved in the production and transportation of manure.[93] Similarly, in western Kenya, the transfer of biomass from trees to land producing crops is constrained both by the availability of materials and the availability of labour.[94]

4.1.2 Water

Agriculture accounts for 80–90% of freshwater use globally so there is considerable focus on using this resource more efficiently.[95,96] Irrigated crops occupy about 15–20% of the total cropped area but contribute 33–40% of the production, so they are crucial to the world's food supply. Most irrigation is applied on the surface (84.5% of the total), with smaller amounts *via* sprinklers (13.5%) and localised systems (2%). Generally, surface irrigation systems have been used to apply large quantities of water, but their on-farm application efficiency is, on average, low, and over-irrigation is common.[95] Globally, storage and conveyancing efficiencies are around 70%, implying a 30% loss of water before delivery to the field.[97] On the farm, irrigation efficiency is typically only about 37%, so that almost two-thirds of the water delivered is lost as drainage or runoff, or both.[95,97] Improving the efficiency of this practice will be critical for future improvement in water-use efficiency (Table 3).

Until recently, most irrigation was scheduled on the basis of fully meeting crop water requirements, but with sprinkler and localised drip systems it has been possible to demonstrate that deficit irrigation strategies can not only sustain yields and profitability, but also reduce water use.[98,99] More developmental work is required to turn these into widespread, commercial systems.

In both irrigated and rain-fed production systems, substantial amounts of water can be lost as evaporation directly from the soil surface, and many

Table 3 Management practices available to increase the efficiency of water use at a field scale in rain-fed and irrigated production systems. (Source: adapted from Wallace).[95]

Improvement type	*Reduce storage and conveyancing losses*	*Reduce runoff and drainage losses*	*Maximise crop transpiration*	*Increase dry matter per unit of water transpired*
Agronomic		Use of fertiliser; using deep rooted varieties; cultivation to improve infiltration	Early sowing and use of varieties with rapid early growth; modifying plant density; mulching; fallowing and weed control	Use of higher-yielding varieties; maximising cropped area during periods of low evaporative demand and/or high rainfall
Technical		Use of practices that increase effectiveness of rainfall; *e.g.* water harvesting, laser levelling, more efficient irrigation methods, such as drip and sub-surface irrigation		
Managerial	Improved maintenance of equipment; better use and management of saline and waste water	Adoption of demand-based irrigation scheduling systems; use of deficit scheduling		
Institutional	User involvement in scheme operation and maintenance; introduction of water pricing and legal frameworks to provide incentives for efficient water use and disincentives for inefficient use; introduction of integrated catchment management; improved training and extension			

workers have commented on the small proportion of water potentially available to crops that is actually transpired in some environments.[71,100] For example, transpiration from barley crops in northern Syria was only a small component (<35%) of total water use.[101] Similarly, in Niger, where rainfall frequently occurs as intensive showers, transpiration was normally less than evaporation from the soil surface,[102] and on sloping land (angle 2–3%) in farmers' fields transpiration was as low as 6% of rainfall.[103] Such findings have led many to conclude that the efficiency with which water is used to produce crops could be significantly improved in many rain-fed environments.[100,104]

Several potential agronomic management options exist for reducing evaporation directly from soil, and increasing the supply of water to crops (Table 3). However, the success or otherwise of a particular management practice in increasing the efficiency with which water is used will depend on a combination of soil and climatic conditions.[105] The physical factors that appear to be important in determining the success of a management option include the moisture characteristic curve and hydraulic conductivity of the soil, the amount of crop cover and the distribution of roots, the quantity and temporal distribution of rain, and the potential rate of evaporation.[105] Future increases in water-use efficiency in rain-fed systems will come, to a considerable extent, from capturing rainfall and then retaining it in soils until it is needed by crops.[106]

4.1.3 Nutrient–Water Interactions

It has been appreciated for a long time that there is an interaction between the efficiencies which water and nutrients are used, not least because the availability and mobility of nutrients in soils depends on water. In rain-fed production systems this interaction is a major constraint to the efficient use of inputs, because rainfall is not known in advance. Where soil nutrients are deficient for maximum growth of crops, application of fertilisers and manures may not only result in increased growth but also in increased water use efficiency (WUE, defined as the quantity of crop dry matter produced per unit of water used). This effect of modest applications of fertiliser has been well documented in several studies[71,107] and is illustrated in Table 4. Fertiliser use may increase slightly the total amount of water used (*e.g.* barley in Syria;[71] maize in the UK),[107] but the principal effect is to increase early canopy growth so that it shades the surface and thereby reduces evaporation from the soil surface as a proportion of the total water that is evaporated. However, the beneficial effect of fertiliser in increasing growth and reducing evaporation from the soil surface is not universal, and is dependent on the wetness of the soil surface and the evaporative demand.[100]

In semi-arid production systems, the efficiency of N and P fertilisers depends on the amount of water available to the crop so that the response to N, in particular, is variable and limited in dry years. Typically, crop response to N increases with increasing rainfall while response to P decreases on P-deficient

Table 4 Effects of modest applications of fertiliser on shoot dry matter, water use and water use efficiency (WUE) for crops of barley at Breda, Syria,[71] and pearl millet at Sadore and Dosso, Niger (ICRISAT). (Source: adapted from Gregory).[100]

Crop	*Season*	*Rainfall (mm)*	*Fertiliser*	*Dry matter ($t\ ha^{-1}$)*	*Water use (mm)*	*WUE ($kg\ ha^{-1}\ mm^{-1}$)*
Barley	1981/82	324	+	6100	231	26.4
			−	4540	231	19.7
	1983/84	204	+	2880	176	16.3
			−	1340	171	7.8
Millet	1984	260	+	4750	165	28.8
			−	2417	163	14.8
	1985	380	+	5000	247	20.2
			−	3100	270	11.5
	1986	440	+	3850	268	14.4
			−	1140	211	5.4

soils.[108] Studies in the Sahel have also concluded that soil fertility is often a more important factor than rainfall in rangeland and crop productivity, so that effective management of water cannot be achieved without also managing soil nutrient constraints.[109] The limiting factors to crop growth at different times during any particular season could be either water or nutrient availability, or both. A practical problem to be resolved in many semi-arid regions is how to afford and apply the optimum amount of fertiliser to produce an economically viable yield in a given season. For example, an analysis of financial returns over a 40-year period from fertiliser applications in an area with low and erratic rainfall in south-eastern Australia found that a very low input of N fertiliser (5 kg N ha^{-1}) ensured the greatest economic stability at all sites examined.[110] Most scientific analysis is conducted with the benefit of hindsight, but farmers must operate without this benefit, so that conservative practices tend to dominate.

4.2 Improved Tillage

In addition to managing chemical and biological elements of soil fertility, there is increasing emphasis on soil physical fertility and health as a means of improving both root growth and function, and soil ecosystem services. The practical means by which these ends can be achieved often relies on tillage. In much conventional agriculture, soil is often inverted (with, for example, a mouldboard plough or discs), with crop residues and weeds being buried. There are many reasons why such tillage practices arose, including: (i) the preparation of a seedbed, allowing uniform germination; (ii) removal of weed competition early in the crop's growing cycle; (iii) release of nutrients through mineralisation after oxidation of soil organic matter; (iv) burial of crop residues, reducing carry-over of diseases and facilitating easier sowing; (v) easier incorporation of nutrients; (vi) relief from soil compaction; and (vii) control of some soil-borne diseases and insects.[111] However, by creating a bare soil surface, such tillage also has the undesirable consequences of substantially increasing rates of water and wind erosion and of the oxidation of soil carbon.

In many regions of the world there has been a move away from such inversion tillage towards more minimal soil disturbance – a move that has been assisted by the development of herbicides to control weeds. Conservation tillage is a wide-ranging term used to describe the many practices that have evolved, initially to retain at least 30% surface coverage by crop residues, but to varying extents to also conserve time, fuel, earthworms, soil water, soil structure and nutrients.[112] Several studies have shown that minimum tillage coupled with residue retention increased infiltration and profile moisture storage (*e.g.* in Kenya),[113] reduced soil erosion by water by 10- to 100-fold because of greater aggregate stability and reduced runoff,[114,115] and increased soil organic matter content.[116]

Conservation tillage has been widely adopted in the USA, Brazil, Argentina, Canada and Australia and now covers over 100 Mha of land worldwide. This tillage and other practices form the basis of "conservation agriculture" which combines three principles of:[117]

- reduced tillage – ideally zero tillage, but may involve controlled tillage during seeding to disturb no more than 20–25 % of the soil surface;
- retention of crop residues and surface cover – variable rates of retention to reduce erosion and runoff and enhance soil properties associated with long-term production; and
- use of crop rotations – moderation of weed, disease and pest problems coupled with utilisation of beneficial effects of some crops on soil conditions.

These principles are applicable to a wide range of production systems from low-yielding, rain-fed conditions to high-yielding, irrigated conditions. In practice, though, the techniques used will vary with the specific situation, depending on farmer circumstances and biophysical and management conditions.

4.3 Improving Root Systems

There is much interest in exploiting genotypic differences in the uptake of nutrients (especially of N and P) and, particularly, in improving the efficiency with which resources are used.[118] The root factors contributing to P-uptake efficiency have been summarised as:[119]

(i) root geometry – differences in root length and its distribution in soils, root-hair length and density, root diameter, *etc.*;
(ii) mycorrhizal effects – differences in the extent or rate of infection, or species of mycorrhizal fungus; and

(iii) solubilisation effects – differences in P solubility close to the root surface arising from changed soil chemical conditions.

This is a complex set of properties that produces inter-related effects on internal, physiological efficiency and the efficiency of recovery (for definitions see ref. 120). The interactions arise mainly because any additional nutrients provided by externally efficient roots may also stimulate root growth. For example, model simulations of rice showed that small changes (22%) in root diameter or internal efficiency had large effects (three-fold) on P uptake. The same result could be achieved by a 33% increase in root external efficiency, but only 10% of the three-fold increase in P uptake was directly attributable to the direct effect of increased external root efficiency, with 90% due to enhanced root growth as a consequence of higher P uptake per unit of root.[121] These studies concluded that large genotypic differences in P uptake from P-deficient soils can result from small differences in tolerance mechanisms and that these small changes will be difficult to detect as changes in recovery efficiency because they are likely to be overshadowed by the effects on root growth.

The importance of different root architecture in response to soil conditions, or as a consequence of genotypic differences in root growth, for nutrient uptake is starting to emerge.[122] For example, four genotypes of common bean representing distinct shoot growth habits (erect determinate, erect indeterminate, prostrate indeterminate, and climbing) were grown in containers of an oxisol and a range of root parameters measured up to 14 days after planting.[123] Table 5 shows that there were significant differences between the genotypes after 14 days in root length and mass, number of roots arising from the base of the hypocotyls (basal roots), and root growth and root elongation rates. The P-efficient genotype Tostado, which grows well in highly acidic, infertile soils in Rwanda, had the most vigorous seedling root system which was highly branched and with numerous basal roots, whereas the landrace Porrillo

Table 5 Root growth parameters for four genotypes of common bean grown in containers of an oxisol for 14 days. Values are the mean of four replicates, with the standard error shown in brackets. (Source: adapted from Lynch).[123]

	Genotype			
	Tostado	*Porrillo sintetico*	*Carioca*	*HAB 229*
Total root dry weight (g plant^{-1})	0.38 (0.03)	0.23 (0.05)	0.27 (0.04)	0.28 (0.01)
Total root length (m plant^{-1})	65.9 (23.9)	23.9 (2.6)	35.1 (5.0)	49.6 (15.6)
Number of basal roots	252 (14)	171 (22)	271 (39)	216 (44)
Relative total root growth rate (d^{-1})	0.20 (0.01)	0.20 (0.04)	0.18 (0.01)	0.25 (0.01)
Relative total root elongation rate (d^{-1})	0.48 (0.03)	0.38 (0.03)	0.40 (0.03)	0.42 (0.02)

sintetico, which grows well on fertile soils in South America, had a smaller, less-branched root system. Such results demonstrate that substantial genetic variability exists for root traits that determine the relative distribution of roots in different soil layers and thereby influence the acquisition of resources. This variation was exploited by crossing deep-rooted and shallow-rooted genotypes of bean to obtain recombinant in-bred lines and it was found that lines with the highest P-acquisition efficiency had shallower root systems.[124] At least two factors are believed to contribute to this greater P efficiency of shallower root systems compared with deeper root systems: first, spatial coincidence of root and resource, and second, lower intra-plant inter-root competition. Such genotypes have the added benefit that their increased growth often results in greater ground cover and less P loss by erosion of topsoil.[125] In soyabean, too, the most P-efficient genotypes had longer and larger root systems with a greater proportion of the root system in the topsoil.[126]

Significant correlations between root architectural features and P uptake and P-use efficiency have also been found in a wide range (355) of *Brassica oleracea* L. accessions. Many measures of P-use efficiency were correlated with root development and architecture, especially with lateral root number, length and growth rate.[127] Physiological P-use efficiency varied four- to five-fold in a range of commercial genotypes, suggesting that there is potential to breed more efficient cultivars. Similarly, a study of two soyabean genotypes, together with their 88 recombinant in-bred lines, found that P-use efficiency was significantly correlated with root length, surface area, root width and root depth.[126] These correlations, together with the high broad-sense heritability values of the root traits, suggest the feasibility of screening P-efficient genotypes through selection of simple root traits in the field.

Root architectural traits are also important in water-limited environments,[128,129] with relations demonstrated between the angular orientation of roots and the subsequent extraction of soil water.[128] There is some evidence that root growth angles in cereals are related to the environment in which the plants evolved. For example, wild and landrace barleys tend to have a narrower angular spread than modern cultivars as a consequence of their evolution in water-limited conditions.[129,130] Modern cultivars tend to have a wider angular spread which allows them to exploit the concentrated topsoil nutrients of fertilised agricultural soils.

Marrying the architectural requirements for soil resources that are mobile, such as water and nitrate, with those that are relatively immobile, such as phosphorus, poses a considerable challenge to plant breeders, but one that is very important if sustainable intensification of crop production is to be achieved.

5 Concluding Remarks

Soils are important capital resources that underpin the functioning of many ecosystem services and especially food production. In the recent past the pressing need to increase production has resulted in many detrimental environmental consequences, but the future challenge will be to increase production while simultaneously minimising other undesirable effects. Achieving food security for all will require investment in research and development and a much greater cross-connection between the various actors in the food system, ranging from farmers to retailers and consumers, together with the use of all appropriate technologies.[84,85]

References

1. FAO, *Report of the World Food Summit*, FAO, Rome, 1996.
2. P. J. Gregory, J. S. I. Ingram and M. Brklacich, *Philos. Trans. R. Soc. London, Ser. B*, 2005, **352**, 2139.
3. P. J. Ericksen, *Global Environ. Change*, 2008, **18**, 234.
4. H. L. Foster, *J. Soil Sci.*, 1981, **32**, 149.
5. M. Latham in *Soil Science and Sustainable Land Management in the Tropics*, ed. J. K. Syers and D. L. Rimmer, CAB International, Wallingford, 1994.
6. N. Ahmad and A. Mermut, *Vertisols and Technologies for Their Management*, Elsevier, Amsterdam, 1996.
7. J. K. Syers, F. W. T. Penning de Vries and P. Nyamudeza, *The Sustainable Management of Vertisols*, CAB International, Wallingford, 2001.
8. P. Marschner, *Mineral Nutrition of Higher Plants*, Academic Press, London, 3rd edn, 2012.
9. N. K. Fageria, V. C. Baligar and C. A. Jones, *Growth and Mineral Nutrition of Field Crops*, CRC Press, Boca Raton, 3rd edn, 2011.
10. P. J. White, M. R. Broadley and P. J. Gregory, *Appl. Environ. Soil Sci.*, 2012, Art. ID 104826.
11. P. J. White and P. H. Brown, *Ann. Bot. (Oxford, U. K.)*, 2010, **105**, 1073.
12. R. D. Graham, R. M. Welch, D. A. Saunders, I. Ortiz-Monasterio, H. E. Bouis, M. Bonierbale, S. de Haan, G. Burgos, G. Thiele, R. Liria, C. A. Meisner, S. E. Beebe, M. J. Potts, M. Kadian, P R. Hobbs, R. K. Gupta, and S. Twomlow, *Adv. Agron.*, 2007, **92**, 1.
13. H. E. Bouis and R. M. Welch, *Crop Sci.*, 2010, **50**, S20.
14. A. J. Stein, *Plant Soil*, 2010, **335**, 133.
15. A. J. Karley and P. J. White, *Curr. Opin. Plant Biol.*, 2009, **12**, 291.
16. P. J. White and M. R. Broadley, *New Phytol.*, 2009, **182**, 49.
17. P. J. White and D. J. Greenwood, in *Soil Conditions and Plant Growth*, ed. P. J. Gregory and S. Nortcliff, Wiley/Blackwell, Oxford, 12th edn, 2012.
18. I. Cakmak, *Plant Soil*, 2008, **302**, 1.
19. M. R. Broadley, P. J. White, J. P. Hammond, I. Zelko and A. Lux, *New Phytol.*, 2007,**173**, 677.
20. S. R. Wilkinson, R. M. Welch, H. F. Mayland and D. L. Grunes, *Metal Ions Biol. Syst.*, 1990, **26**, 33.
21. G. H. Lyons, J. C. R. Stangoulis and R. D. Graham, *Nutr. Rev.*, 2004, **62**, 247.

22. J. F. Risher and L. S. Keith, *Iodine and Inorganic Iodides: Human Health Aspects*, WHO Press, Geneva, 2009.
23. H. Hartikainen, *J. Trace Elements Med. Biol.*, 2005, **18**, 309.
24. M. R. Broadley, P. J. White, R. J. Bryson, M. C. Meacham, H. C. Bowen, S. E. Johnson, M. J. Hawkesford, S. P. McGrath, F. -J. Zhao, N. Breward, M. Harriman and M. Tucker, *Proc. Nutr. Soc.*, 2006, **65**, 169.
25. M. E. Hodson and E. Donner, in *Soil Conditions and Plant Growth* , ed. P. J. Gregory and S. Nortcliff, Wiley/Blackwell, Oxford, 12th edn, 2012.
26. C. van Leeuwen, P. Friant, X. Choné, O. Tregoat, S. Koundouras and D. Dubourdieu, *Am. J. Enol. Vitic.*, 2004, **55**, 207.
27. D. J. Greenland, *The Sustainability of Rice Farming*, CAB International, Wallingford, 1997.
28. D. L. Rowell, *Soil Science: Methods and Applications*, Prentice Hall, London, 1994.
29. E. A. Stockdale, K. W. T. Goulding, T. S. George and D. V. Murphy, in *Soil Conditions and Plant Growth*, ed. P.J. Gregory and S. Nortcliff, Wiley/Blackwell, Oxford, 12th edn, 2012.
30. C. J. M. Konsten, N. van Breemen, S. Suping, I. B. Aribawa and J. E. Groenenberg, *Soil Sci. Soc. Am. J.*, 1993, **58**, 871.
31. D. C. Edmeades and A. M. Ridley, in *Handbook of Soil Acidity*, ed. Z. Rengel, Marcel Dekker Inc., New York, p. 297.
32. A. S. Lopes, *Better Crops Int.*, 1996, **10**, 9.
33. A. Murányi, B. Seeling, E. Ladevig and A. Jungk, *J. Plant Nutr. Soil Sci.*, 1994, **157**, 61.
34. C. Tang, M. J. Unkovich and J. W. Bowden, *New Phytol.*, 1999, **143**, 513.
35. FAO, *World Reference Base for Soil Resources*, FAO, Rome, 1998.
36. A. Wild, *Soils, Land and Food: Managing the Land during the Twenty-First Century*, Cambridge University Press, Cambridge, 2003.
37. P. Rengasamy, *J. Exp. Bot.*, 2006, **57**, 1017.
38. Western Australian Department of Agriculture, *Wheatbelt Salinity: A Review of the Salt Land Problem in South-Western Australia*, Tech. Bull. 52, 1983.
39. R. Munns and M. Tester, *Ann. Rev. Plant Biol.*, 2008, **59**, 651.
40. R. Munns, New Phytol., 2005, **167**, 645.
41. Y. Genc, G. K. McDonald and M. Tester, *Plant Cell Environ.*, 2007, **30**, 1486.
42. S. J. Roy, E. J. Tucker and M. Tester, *Curr. Opin. Plant Biol.*, 2011, **14**, 232.
43. P. D. Hallett and A. G. Bengough, in *Soil Conditions and Plant Growth*, ed. P. J. Gregory and S. Nortcliff, Wiley/Blackwell, Oxford, 12th edn, 2012.
44. A. R. Dexter, *Soil Tillage Res.*, 1988, **11**, 199.
45. P. J. Gregory, A. G. Bengough, T. S. George and P. D. Hallett, in *Advances in Modeling Agricultural Systems: Trans-disciplinary Research, Synthesis, Modeling, and Applications. Vol 3. Enhancing Understanding and Quantification of Soil-Root Growth Interactions*, ed. D. Timlin and L. Ahuja, ASA-SSSA-CSSA, Madison, WI, 2012.
46. M. Watt, M. E. McCully and M. J. Canny, *Plant Physiol.*, 1994, **106**, 179.

47. F. Watteau, G. Villemin, G. Burtin and L. Jocteur-Monrozier, *Eur.. J. Soil Sci.*, 2006, **57**, 247.
48. S. Czarnes, A.R. Dexter, P. D. Hallett and F. Bartoli, *Plant Soil*, 1999, **211**, 69.
49. A. Pierret, C. Doussan, Y. Capowiez, F. Bastardie and L. Pages, *Vadose Zone J.*, 2007, **6**, 269.
50. H. el Zemrany, S. Czarnes, P. D. Hallett, S. Alamercery, R. Bally and L. Jocteur-Monrozier, *Plant Soil*, 2007, **291**,109.
51. P. D. Hallett, D. S. Feeney, A. G. Bengough, M. C. Rillig, C. M. Scrimgeour and I. M. Young, *Plant Soil*, 2009, **314**, 183.
52. S. Yoshida and P. D. Hallett, *Water Resour. Res.*, 2008, **44**, WC001.
53. R. G. White and J. A. Kirkegaard, *Plant Cell Environ.*, 2020, **33**, 133.
54. I. M. Young, *J. Agric. Sci.*, 1998, **130**, 1.
55. A. Vollsnes, C. Futsaether and A. G. Bengough, *Eur. J. Soil Sci.*, 2010, **48**, 926.
56. B. W. Veen, M. Noordwijk, P. Willigen and F. R. Boone, *Plant Soil*, 1992, **139**, 131.
57. S. Schmidt, A. G. Bengough, P. J. Gregory, D. V. Grinev and W. Otten, *Eur. J. Soil Sci.*, 2013, in press.
58. H. C. J. Godfray, J. R. Beddington, I. R. Crute, L. Haddad, D. Lawrence, J. F. Muir, J. Pretty, S. Robinson, S. M. Thomas and C. Toulmin, *Science*, 2010, **327**, 812.
59. P. Smith, P. J. Gregory, D. van Vuuren, M. Obersteiner, P. Havlik and M. Rounsvell, *Philos. Trans. R. Soc. London, Ser. B*, 2010, **365**, 2941.
60. P. J. Gregory, J. S. I. Ingram, R. Andersson, R. A. Betts, V. Brovkin, T. N. Chase, P. R. Grace, A. J. Gray, N. Hamilton, T. B. Hardy, S. M. Howden, A. Jenkins, M. Meybeck, M. Olsson, I. Ortiz-Monasterio, C. A. Palm, T.W. Payn, M. Rummukainen, R. E. Schulze, M. Thiem, C. Velentin and M. J. Wilkinson, *Agric. Ecosyst. Environ.*, 2002, **88**, 279.
61. J. Bruinsma, *World Agriculture: Towards 2015/2030, an FAO Perspective*, Earthscan Publications, London, 2003.
62. R. K. M. Hay and J. R. Porter, *The Physiology of Crop Yield, Second Edition*, Blackwell, Oxford, 2008.
63. L. T. Evans, *Crop Evolution, Adaptation and Yield*, Cambridge University Press, Cambridge, 1993.
64. L. T. Evans, *Feeding the Ten Billion; Plants and Population Growth*, Cambridge University Press, Cambridge, 1998.
65. F. Ewert, M. D. A. Rounsvell, I. Reginster, M. J. Metzger and R. Leemans, *Agric Ecosyst. Environ.*, 2005, **107**, 101.
66. J. F. Angus, *Aust. J. Exp. Agric.*, 2001, **41**, 277.
67. P. A. Sanchez, K. D. Shepherd, M. J. Soule, F. M. Place, R. J. Buresh, A. -M. N. Izac, A. U. Mokwunye, F. R. Fwesiga, C. G. Ndiritu and P. L. Woomer, in *Replenishing Soil Fertility in Africa*, ed. R. K. Buresh, P. A. Sanchez and F. Calhoun, Soil Science Society of America, Madison, 1997, p 1.

68. D. J. Greenland, P. J. Gregory and P. H. Nye, in *Feeding a World Population of More Than Eight Billion People: A Challenge to Science*, ed. J. C. Waterlow, D. G. Armstrong, L. Fowden and R. Riley, Oxford University Press, Oxford, 1998, p. 39.
69. K. W. Jaggard, A. Qi and E. S. Ober, *Philos. Trans. R. Soc. London, Ser. B*, 2010, **365**, 2835.
70. D. B. Lobell, K. G. Cassman and C. B. Field, *Ann. Rev. Environ. Res.*, 2009, **34**, 179.
71. P. J. M. Cooper, P. J. Gregory, D. Tully and H. C. Harris, *Exp. Agric.* 1987, **23**, 113.
72. K. Neumann, P. H. Verburg, E. Stehfest and C. Müller, *Agric. Syst.*, 2010, **103**, 316.
73. X. -P. Chen, Z. -L. Cui, P. M. Vitousek, K. G. Cassman, P. A. Matson, J. -S. Bai, Q. -F. Meng, P. Hou, S. -C. Yue, V. Römheld and F. -S. Zhang, *Proc. Natl. Acad. Sci.* U. S. A., 2011, **108**, 6399.
74. A. G. Laborte, K. de Bie, E. M. A. Smaling, P. F. Moya, A. A. Boling and M. K. van Ittersum, *Eur. J. Agron.*, 2012, **36**, 9.
75. P. M. Vitousek, R. Naylor, T. Crews, M. B. David, L. E. Drinkwater, E. Holland, P. J. Johnes, J. Katzenberger, L. A. Martinelli, P. A. Matson, G. Nziguheba, D. Ojima, C. A. Palm, G. P. Robertson, P. A. Sanchez, A. R. Townsend and F. S. Zhang, *Science*, 2009, **324**, 1519.
76. J. N. Quinton, G. Govers, K. van Oost and R. D. Bardgett, *Nature Geosci.*, 2010, **3**, 311.
77. MEA, *Millenium Ecosystem Assessment*, Island Press, Washington DC, 2005.
78. J. Pretty, *Philos. Trans. R. Soc. London, Ser. B*, 2008, **363**, 447.
79. Foresight, *The Future of Food and Farming*, Government Office of Science, London, 2011.
80. D. S. Powlson, P. J. Gregory, W. R. Whalley, J. N. Quinton, D. W. Hopkins, A. P. Whitmore, P. R. Hirsch and K. W. T. Goulding, *Food Policy*, 2011, **36**, S72.
81. R. Lal, *Science*, 2004, **304**, 1623.
82. K. G. Cassman, A. Dobermann, D. T. Walters and H. Yang, *Ann. Rev. Environ. Res.*, 2003, **28**, 315.
83. R. M. Ewers, J. P. W. Scharlemann, A. Balmford and R. E. Green, *Global Change Biol.*, 2009, **15**, 1716.
84. N. Koning and M. K. van Ittersum, *Curr. Opin. Environ. Sustainability*, 2009, **1**, 77.
85. A. Misselhorn, P. Aggarwal, P. Ericksen, P. Gregory, L. Horn-Phathanothai, J. Ingram and K. Wiebe, *Curr. Opin. Environ. Sustainability*, 2012, **4**, 7.
86. D. Tilman, C. Balzer, J. Hill and B. L. Befort, *Proc. Natl. Acad. Sci.* U. S. A., 2011, **108**, 20260.
87. R. J. K. Myers, C. A. Palm, E. Cuevas, I. U. N. Gunatilleke and M. Brossard, in *The Biological Management of Tropical Soil Fertility*, ed. P. L. Woomer and M. J. Swift, John Wiley & Sons, Chichester, 1994, p. 81.

88. S. McCormick, C. Jordan and J. S. Bailey, *Precision Agric.*, 2009, **3**, 262.
89. R. A. Lawes and M. J. Robertson, *Field Crops Res.*, 2011, **124**, 142.
90. M. J. Robertson, G. Lyle and J. W. Bowden, *Field Crops Res.*, 2008, **105**, 211.
91. D. P. Sherchan, C. J. Pilbeam and P. J. Gregory, *Exp. Agric.*, 1999, **35**, 1.
92. C. Turton, A. Vaidya, J. K. Tuladhar and K. D. Joshi, *Towards Sustainable Soil Fertility Management in the Hills of Nepal*, Natural Resources Institute, Chatham Maritime, 1995.
93. C. J. Pilbeam, B. P. Tripathi, R. C. Munankarmy and P. J. Gregory, *Mountain Res. Develop.*, 1999, **19**, 333.
94. G. Shepherd, R. J. Buresh and P. J. Gregory, *Biol. Fertility Soils*, 2000, **31**, 348.
95. J. S. Wallace and P. J. Gregory, *Aquatic Sci.*, 2002, **64**, 363.
96. J. I. L. Morison, N. R. Baker, P. M. Mullineaux and W. J. Davies, *Philos. Trans. R. Soc. London, Ser. B*, 2008, **363**, 639.
97. J. S. Wallace, *Agric. Ecosyst. Environ.*, 2000, **82**, 105.
98. E. Fereres and M. A. Soriano, *J. Exp. Bot.*, 2007, **58**, 147.
99. I. C. Dodd, *J. Exp. Bot.*, 2009, **60**, 2454.
100. P. J. Gregory, L. P. Simmonds and G. P. Warren, *Philos. Trans. R. Soc. London, Ser. B*, 1997, **352**, 987.
101. S. J. Allen, *Agric. Forestry Meteorol.*, 1990, **49**, 291.
102. J. S. Wallace, C. R. Lloyd and M. V. K. Sivakumar, *Agric. Forestry Meteorol.*, 1993, **63**, 149.
103. J. Rockström, On-farm agrohydrological analysis of the Sahelian yield crisis: rainfall partitioning, soil nutrients and water use efficiency of pearl millet, PhD Thesis, University of Stockholm, 1997.
104. W. A. Payne, *Agron. J.*, 1997, **89**, 481.
105. P. J. Gregory, L. P. Simmonds and C. J. Pilbeam, *Agron. J.*, 2000, **92**, 814.
106. W. A. Payne, in *Food Security and Soil Quality*, ed. R. Lal and B. A. Stewart, CRC Press, Boca Raton, 2010, p. 23.
107. J. B. O. Ogola, T. R. Wheeler and P. M. Harris, *Field Crops Res.*, 2002, **78**, 105.
108. M. J. Jones and A. Wahbi, *Exp. Agric.*, 1992, **28**, 63.
109. W. A. Payne, *Agron. J.*, 2000, **92**, 808.
110. V. Sadras, *Field Crops Res.*, 2002, **77**, 201.
111. P. R. Hobbs, K. Sayre and R. Gupta, *Philos. Trans. R. Soc. London, Ser. B*, 2008, **363**, 543.
112. C. J. Baker, K. E. Saxton and W. R. Ritchie, *No-Tillage Seeding: Science and Practice*, CAB International, Wallingford, 2nd edn, 2006.
113. P. T. Gicheru, *Soil Technol.*, 1994, **7**, 209.
114. F. E. Rhoton, M. J. Shipitalo and D. L. Lindbo, *Soil Tillage Res.*, 2002, **66**, 1.
115. D. R. Montgomery, *Proc. Natl. Acad. Sci. U. S. A.*, 2007, **104**, 13268.
116. B. Govaerts, N. Verhulst, K. D. Sayre, J. Dixon and L. Dendooven, *Crit. Rev. Plant Sci.*, 2009, **28**, 97.

117. N. Verhulst, B. Govaerts, E. Verachtert, A. Castellanos-Navarrete, M. Mezzalama, P. C. Wall, A. Chocobar, J. Deckers and K. D. Sayre in *Food Security and Soil Quality*, ed. R. Lal and B. A. Stewart, CRC Press, Boca Raton, 2010, p. 137.
118. P. J. Gregory, *Plant Roots: Growth, Activity and Interactions with Soil*, Blackwells, Oxford, 2006.
119. G. J. D. Kirk, T. George, B. Courtois and D. Senadhira, *Field Crops Res.*, 1998, **56**, 73.
120. J. K. Ladha, H. Pathak, T. J. Krupnik, J. Six and C. van Kessel, *Adv. Agron.*, 2005, **87**, 85.
121. M. Wissuwa, *Plant Physiol.*, 2003, **133**, 1947.
122. J. P. Lynch, *Aust. J. Bot.*, 2007, **55**, 1.
123. J. P. Lynch and J. J. van Beem, *Crop Sci.*, 1993, **33**, 1253.
124. H. Liao, X. Yan, G. Rubio, S. E. Beebe, M. W. Blair and J. P. Lynch, *Funct. Plant Biol.*, 2004, **31**, 959.
125. A. Henry, N. F. Chaves, P. J. A. Kleinman and J. P. Lynch, *Field Crops Res.*, 2010, **115**, 67.
126. J. Ao, J. Fu, J. Tian, X. Yan and H. Liao, *Funct. Plant Biol.*, 2010, **37**, 304.
127. J. P. Hammond, M. R. Broadley, P. J. White, G. J. King, H. C. Bowen, R. Hayden, M. C. Meacham, A. Mead, T. Overs, W. P. Spracklen and D. J. Greenwood, *J. Exp. Bot.*, 2009, **60**, 1953.
128. A. M. Manschadi, J. Christopher, P. deVoil and G. L. Hammer, *Funct. Plant Bi*ol., 2006, **33**, 823.
129. C. E. Hargreaves, P. J. Gregory and A. G. Bengough, *Plant Soil*, 2009, **316**, 285.
130. A. G. Bengough, D. C. Gordon, H. Al-Menaie, R. P. Ellis, D. Allan, R. Keith, W. T. B. Thomas and B. P. Forster, *Plant Soil*, 2004, **262**, 63.

Global Soils: Preserving the Capacity for Food Production

LUCA MONTANARELLA

European Commission - DG JRC, Via E. Fermi, 2749, I-21027 Ispra (VA), Italy
E-mail: luca.montanarella@jrc.ec.europa.eu
http://eusoils.jrc.it/index.html
http://esdac.jrc.ec.europa.eu/

ABSTRACT

Global soil resources are limited and non-renewable. Therefore there is the need to establish a framework for preserving these limited resources for future generations. Current rates of soil degradation and land-take by urban sprawl and other land uses is severely threatening the capacity of planet Earth to produce sufficient food for the projected population of more than 9 billion by 2050. A clear target needs to be set and a functioning global governance system needs to be established for halting the rate of soil loss in order to preserve this precious resource. After 20 years from the Rio Convention, a binding target of zero net soil degradation by 2050 needs to be approved by all Nations as a fundamental sustainable development goal, together with a global soil partnership for soil protection federating all relevant stakeholders towards the achievement of this ambitious goal.

Issues in Environmental Science and Technology No. 35
Soils and Food Security
Edited by R.E. Hester and R.M. Harrison

Published by the Royal Society of Chemistry, www.rsc.org

1 Introduction

Sustainable development is development that meets the needs of the present without compromising the ability of future generations to meet their own needs.[1] Given the current population growth trends and the forecasted global population of more than 9.3 billion by 2050,[2] it seems a rather ambitious target to achieve. Non-renewable natural resources are becoming depleted at a rate that will certainly not allow future generations to meet their own needs, unless we adopt a new approach to the management of these resources. Sources of minerals, metals and energy, as well as stocks of fish, timber, water, fertile soils, clean air, biomass, and biodiversity are all under pressure, as is the stability of the climate system. Whilst demand for food, feed and fibre may increase by 70% by 2050, 60% of the world's major ecosystems that help produce these resources have already been degraded or are used unsustainably. If we carry on using resources at the current rate, by 2050 we will need, on aggregate, the equivalent of more than two planets to sustain us, and the aspirations of many for a better quality of life will not be achieved.[3]

A fundamental issue that is emerging as the main concern for sustainable development is the key question of food security for the 9 billion (or maybe more) humans that will populate the planet earth by 2050. Will there be enough food for all? We know that already now we are not able to feed the world's population: 850 million people in the world are still suffering from undernourishment and there is no sign of improvement of this figure.[4] Recent assessments[5] indicate that addressing the challenge of future food security is a complex issue that needs to take into account a much broader perspective than in the past, integrating the various socio-economic factors with the available knowledge in agricultural sciences and technologies. Nevertheless, it remains a basic fact that without sufficient soil and water resources[6] any policy intervention and adjustment will be of little effect in solving the food security problem at the global scale.

Preserving the Earth's lands and soils is crucial if we are to provide sufficient food, clean water and healthy recreational spaces, and cut greenhouse gas emissions. We need to use land and soil resources more sustainably, set a measurable path towards preventing degradation, and strengthen existing global governance to tackle land and soil degradation.

Land and soil are finite resources. The growth in world population, the rising consumption of meat and dairy products, and the increased use of biomass for energy and other industrial purposes, all lead to additional pressure on land and soil worldwide and a shrinking land base.

Soils are not only the basis for food production, but also for delivering numerous ecosystems services and functions relevant to human well-being, like water storage, filtration and buffering.

Soil is important for mitigating climate change and its management can support human adaptation efforts. Soil is at the same time both a source of and a sink for greenhouse gases. There is a delicate balance between sink and source functions. Soil contains, worldwide, twice as much carbon as the atmosphere. The flux of carbon dioxide between soil and the atmosphere is also large and estimated at ten times the flux of carbon dioxide from fossil fuels. Waterlogged and permafrost soils hold major stocks of carbon but also are important emitters of two other non-CO_2 greenhouse gases: methane and nitrous oxide.

2 Global Distribution of Soil Resources *versus* Food Production

The area of soil in agriculture currently stands at approximately 4600 million ha. Of this, around 1400–1600 million hectares of soil is cultivated for crops. This area has remained essentially stable over the past 30 years, with a marginal increase of only 8%. Given the constant increase of world population, we have been observing a constant decrease of available cropland per person, which now has fallen below 0.25 ha per person. At the same time, we have been observing a dramatic increase of land-take by other land uses competing with food production, especially in developing countries, where millions of hectares of prime cropland are lost to housing and infrastructure.[7] Additionally, we are losing fertile soils due to various soil degradation processes that are still on-going and do not show any sign of improvement. According to the United Nations Environment Programme (UNEP),[8] 20–50 000 km^2 are lost annually through land degradation, mainly due to soil erosion, with losses two-to-six times higher in Africa, Latin America and Asia than in North America and Europe. Erosion rates in Africa can reach up to 100 tonnes per ha per yr, while wind erosion is a major problem in West Asia, with as much as one-third of the region affected. Each year, the planet loses 24 billion tonnes of topsoil. Over the last two decades, enough has been lost to cover the entire cropland of the United States.

Global soil resources are limited and non-renewable, if not in geological timeframes. Therefore there is the need to manage these limited resources in a sustainable manner in order to ensure that sufficient fertile soils will still be available for future generations. Sustainable soil management requires the adoption of legal frameworks that enforce the implementation of good practices for soil protection by landowners and major stakeholders.[9] Successful examples of national soil protection strategies and legislation exist.[10] Probably the best known example is the US Soil Protection Act of 1935, reversing the dramatic soil degradation processes occurring at that time in the Midwest of USA ("the Dust Bowl") causing extensive erosion by wind and water.[11] Nevertheless, in today's globalised world, single national soil protection strategies are often difficult to implement, given the current pressure on the limited soil resources and the interlinkages between the various socio-economic factors driving land use and land use changes that go far beyond national borders. Indirect land use changes are well documented, especially in conjunction with recently established biofuel production targets by developed countries.[12] The consumption of global cropland (domestic production plus imports minus exports) for the European Union (EU) in 2007, when biofuel production was still very low, was 0.31 hectares per person. This is a third more than the average number of hectares available per world citizen (0.23 hectares).[13] Extensive processes of "land grabbing"[14] by countries seeking a larger fertile soil base for domestic food requirements has underlined the need for global governance of this limited resource. The potential conflicts that may rise in the near future for the stiff competition for soils, as well as for water, requires global governance instruments that guarantee a sustainable management of these natural resources for the benefit of all.

3 Towards Global Governance of Soil Resources

There has been a series of attempts in the past to develop global governance instruments for soils. The first attempt was initiated by the UN Food and Agriculture Organisation (FAO) in 1982 with the adoption of the World Soil Charter,[15] spelling out thirteen basic principles to be adopted by FAO Member States for sustainable soil management and protection (see Box 1). Unfortunately, despite all the efforts by FAO, only little impact could be observed by this initiative on the actual situation on the ground. Nevertheless, the principles of the World Soil Charter remain valid and should continue to guide our actions today for sustainable soil management.

A renewed interest in global soil governance and legal frameworks emerged again in the late 1990s with the proposal of a Convention on the Sustainable Use of Soils.[16] This proposal, elaborated by the Protestant Academy of Tutzing, Germany, was discussed extensively at many stakeholder meetings and conventions[17] but never gained the political consensus needed for entering the intergovernmental debate.

Only recently, following the 2008 food crisis (see Figure 1), a recognition that soil resources for food production are limited has been emerging among policymakers. As a consequence, a surge of interest in the full assessment of available global soil resources has started to become apparent, with major projects aiming towards the collection of updated data and information about soils at detailed scales.[18] Available data already demonstrate that degradation trends of soil resources are increasing and that the global amount of fertile soils is rapidly shrinking.[6]

The FAO, with the support of the European Commission, launched the Global Soil Partnership (GSP) in September 2011 to raise awareness of decision-makers on the vital role of soil resources for achieving food security, for adapting to and mitigating climate change, and guaranteeing the provision of environmental services.

Maintaining healthy soils required for feeding the growing population of the world and meeting their needs for biomass (energy), fibre, fodder and other products can only be ensured through a strong partnership. This is one of the key guiding principles of the GSP, in addition to maintaining soil for other essential ecosystem services on which humans depend for water regulation and

Box 1: Principles of the World Soil Charter

1. *Among the major resources available to men and women is land, comprising soil, water and associated plants and animals: the use of these resources should not cause their degradation or destruction because humans' existence depends on their continued productivity.*
2. *Recognising the paramount importance of land resources for the survival and welfare of people and economic independence of countries, and also the rapidly increasing need for more food production, it is imperative to give high priority to promoting optimum land use, to maintaining and improving soil productivity and to conserving soil resources.*
3. *Soil degradation means partial or total loss of productivity from the soil, either quantitatively, qualitatively, or both, as a result of such processes as soil erosion by water or wind, salinisation, waterlogging, depletion of plant nutrients, deterioration of soil structure, desertification and pollution. In addition, significant areas of soil are lost daily to non-agricultural uses. These developments are alarming in the light of the urgent need for increasing production of food, fibers and wood.*
4. *Soil degradation directly affects agriculture and forestry by diminishing yields and upsetting water regimes, but other sectors of the economy and the environment as a whole, including industry and commerce, are often seriously affected as well through, for example, floods or the silting up of rivers, dams and ports.*
5. *It is a major responsibility of governments that land-use programmes include measures toward the best possible use of the land, ensuring long-term maintenance and improvement of its productivity, and avoiding losses of productive soil. The land users themselves should be involved, thereby ensuring that all resources available are utilised in the most rational way.*
6. *The provision of proper incentives at farm level and a sound technical, institutional and legal framework, are basic conditions to achieve good land use.*
7. *Assistance given to male and female farmers and other land users should be of a practical service-oriented nature and should encourage the adoption of measures of good land husbandry.*
8. *Certain land-tenure structures may constitute an obstacle to the adoption of sound soil management and conservation measures on farms. Ways and means should be pursued to overcome such obstacles with respect to the rights, duties and responsibilities of land owners, tenants and land users alike, in accordance with the recommendations of the Voluntary Guidelines on the Responsible Governance of Tenure of Land and Other Natural Resources (Rome, 2011).*
9. *Land users and the broad public should be well informed of the need and the means of improving soil productivity and conservation. Particular*

Box 1: (Continued)

emphasis should be placed on education and extension programmes and training of agricultural staff at all levels.

10. *In order to ensure optimum land use, it is important that a country's land resources be assessed in terms of their suitability at different levels of inputs for different types of land use, including agriculture, grazing and forestry.*
11. *Land having the potential for a wide range of uses should be kept in flexible forms of use so that future options for other potential uses are not denied for a long period of time or forever. The use of land for non-agricultural purposes should be organised in such a way as to avoid, as much as possible, the occupation or permanent degradation of good-quality soils.*
12. *Decisions about the use and management of land and its resources should favour the long-term advantage rather than the short-term expedience that may lead to exploitation, degradation and possible destruction of soil resources.*
13. *Land conservation measures should be included in land development at the planning stage and the costs included in development planning budgets.*

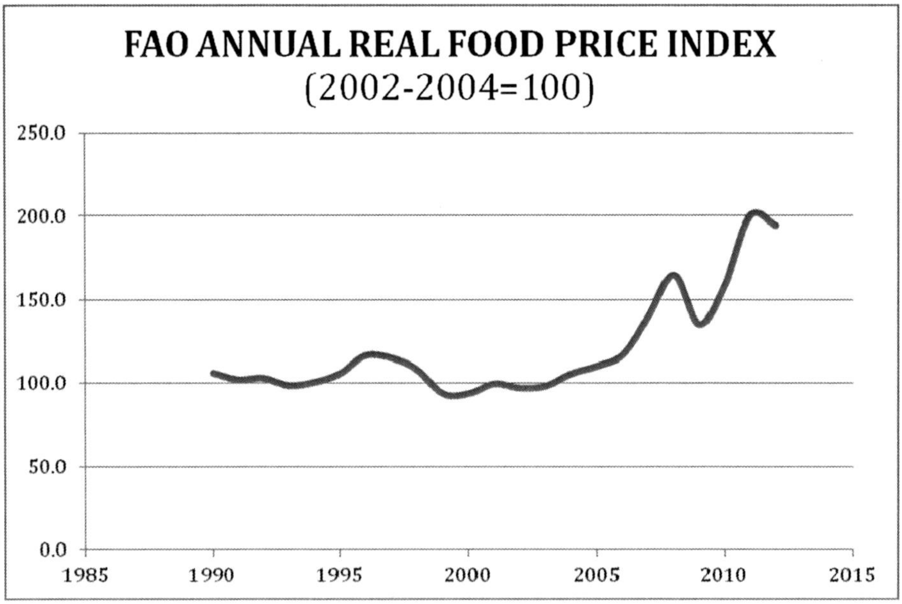

Figure 1 FAO Annual Food Price Index (deflated using the World Bank Manufactures Unit Value Index (MUV) rebased from 1990 = 100 to 2002–2004 = 100) showing the peak food prices on world markets in 2008. (Data source: FAO).

supply, climate regulation, biodiversity conservation and cultural services. The conservation and, where possible, enhancement and restoration of world soil resources through sustainable and productive use should therefore be the ultimate twinned goals of the GSP.

The vision of the Global Soil Partnership is to improve global governance of the limited soil resources of the planet in order to guarantee healthy and productive soils for a food secure world, as well as sustain other essential ecosystem services on which our livelihoods and societies depend, including water regulation and supply, climate regulation, biodiversity conservation and cultural services.

The mission of the GSP is to develop capacities, build on best available science, and facilitate/contribute to the exchange of knowledge and technologies among stakeholders, existing multilateral environmental agreements, and technical and scientific bodies of a similar nature. The GSP will support sustainable management of soil resources at all levels, with a view to enhancing food security, protecting ecosystem services and, in this way, contributing to poverty alleviation in an era of global demographic growth and unsustainable consumption patterns.

Through enhanced and applied knowledge of soil resources as well as improved global governance and standardisation, the Partnership will:

- Create and promote awareness among decision makers and stakeholders of the key role of soil resources for sustainable and productive land management and sustainable development;
- Address critical soil issues aiming at increasing food security and climate change adaptation and mitigation;
- Guide soil knowledge management and targeted research to address concrete challenges on the ground through a common global communication platform;
- Establish an active and effective network for addressing soil cross-cutting issues and ensuring synergies among relevant agricultural and environmental processes; and
- Develop global governance guidelines aiming to improve soil protection and its management for sustainable soil productivity.

The Global Soil Partnership should be based on regional soil partnerships deeply rooted into the national and local stakeholder communities. The partnership is open to all stakeholders that join in a common effort to manage available soil resources in a sustainable way by adopting the principles of the World Soil Charter. Technical and scientific guidance on the partnership will be provided by an Intergovernmental Technical Panel on Soils (ITPS), composed of high-level technical and scientific experts on soil-related issues. This panel will provide institutional and thematic advice to the GSP Secretariat, including the mainstreaming of soils and land-use issues and solutions into the wider regional and international processes, interventions that address the integrated planning and management of land resources, and the achievement of the Millennium Development Goals, particularly supporting the World Food Summit (WFS) Plan of Action and the UN Convention to Combat Desertification (UNCCD), the UN Framework Convention on Climate Change (UNFCCC) and the Convention on Biological Diversity (CBD) in soil-related issues. It will provide the urgently needed science-policy platform for land, as advocated during a recent stakeholder survey.[19] It will complement similar scientific advisory panels for climate change (the Intergovernmental Panel on Climate Change, IPCC) and biodiversity (the Intergovernmental Platform on Biodiversity and Ecosystem Services, IPBES) by providing the needed soil-related data, information and assessments relevant to the various policy-making processes. Particularly in relation to climate change, it will address the urgent need to preserve the available soil organic carbon pool,[20] whilst for biodiversity it will address the neglected aspects of below-ground biodiversity.[21]

It is proposed that the GSP should address five main pillars of action:

- Promote sustainable management of soil resources for soil protection, conservation and sustainable productivity;
- Encourage investment, technical cooperation, policy, education awareness and extension in soils;
- Promote targeted soil research and development, focusing on identified gaps and priorities and synergies with related productive, environmental and social development actions;
- Enhance the quantity and quality of soil data and information: data collection (generation), analysis, validation, reporting, monitoring and integration with other disciplines; and
- Harmonise methods, measurements and indicators for the sustainable management and protection of soil resources.

The Global Soil Partnership is expected to provide the necessary framework towards the long-term goal of achieving a substantial reduction of soil degradation on a global scale. As a possible target, it is proposed[22] to aim towards "Zero Net Land Degradation" by 2030. Achieving such an ambitious target will not only need substantial efforts towards global governance of soil resources, but will also need fundamental research for the definition of measurable indicators and data for assessing the progress made in reducing land and soil degradation.

4 Towards Zero Net Soil Degradation

There is a need for clearly defined targets and time horizons for achieving sustainability of the current economic, social and environmental policies if we want to be able to live on a single planet Earth in peace and prosperity. Preserving the available land for delivering the necessary ecosystem services for all of us requires a careful management of the still-available fertile soils in the world. Certainly setting a target of "zero net land degradation", as advocated by the UNCCD, is important, but needs to be further detailed in order to make it applicable and understandable by all stakeholders and decision makers. There are differences between urban expansion on bare deserts, as, for example, in some of the countries in the Gulf of Persia area (Dubai, Doha, *etc*...), and the rapid urban expansion on prime fertile soils of the Loess Plateau of China. We need to distinguish between soil properties and the related functions that these soils can perform for us. A target of "zero net soil degradation" needs to be seen in relation to the functions certain soils perform. Soils with the primary function of food production, like the highly fertile Chernozems, need to be preserved for that function; therefore, we should aim towards a target of "zero net soil fertility loss" when we address the issue of maintaining the necessary food-producing capacity on planet Earth. In the same manner we need to target "zero net carbon loss" if we address the fundamental function of soils to act as a carbon sink, thus mitigating climate change. Setting clear targets will therefore require a clear classification of the various soil types according to their main functions and the ecosystem services they deliver. A first attempt at such a re-definition of soil quality is on-going[23] and will lead to a newly defined composite indicator, allowing for the clear and measurable assessment of the global target of "zero net soil degradation".

5 Conclusions

A green economy and improved governance for sustainable development should be the main outcomes of the Rio+20 Conference in 2012. The role of soil resources in this context has been well recognised by stakeholders and policymakers. The proposed Global Soil Partnership should provide the future framework for the sustainable use of soil resources for food security and climate change adaptation and mitigation. After several failed attempts in the past, this might be the right moment for an effective step towards soil protection in the world, hopefully delivering the necessary improvement of our environment and of livelihoods, especially in developing countries. Lastly, the fight against land and soil degradation should be part of potential sustainable development goals (SGDs), were they to be agreed at the Conference. This should include setting out a clear path towards minimising land and soil degradation, and it should take into account key processes such as erosion, the loss of soil organic matter, and the disappearance of productive land through urban sprawl.

References

1. WCED, *Our Common Future*, Oxford University Press, 1987.
2. United Nations, *World Population Prospects: The 2010 Revision*, United Nations Department of Economic and Social Affairs (DESA), New York, 2011.
3. European Commission, *Roadmap to a Resource Efficient Europe*, COM(2011) 571, Brussels, 2011.

4. FAO, *The State of Food Insecurity in the World.* FAO, Rome, 2011.
5. The Government Office for Science, Foresight, *The Future of Food and Farming*, Final Project Report, London, 2011.
6. FAO, *The State of the World's Land and Water Resources for Food and Agriculture*, FAO, Rome, 2011.
7. European Commission, *Report on Best Practices for Limiting Soil Sealing and Mitigating its Effects*, Technical Report- 2011 – 050, Brussels, 2011.
8. UNEP, *Global Environmental Outlook*, GEO-4, 2007.
9. I. Hannam and B. Boer, *Drafting Legislation for Sustainable Use of Soils: A Guide*, IUCN, 2004.
10. L. Montanarella, R. Arnold and E. Micheli, Soil Conservation Services in the European Union and in the United States of America, in *Advances of Geoecology*, 2009, **40**, 399.
11. D. Helms, in *A Historical Guide to the U.S. Government*, Oxford University Press, 1998, pp. 434–439.
12. International Council for Science, *Rapid Assessment on Biofuels and the Environment: Overview and Key Findings*, The Scientific Committee on Problems of the Environment (SCOPE), 2009.
13. S. Bringezu, M. O'Brien and H. Schütz, *Land Use Policy*, 2012, **29**, 224–232.
14. L. Cotula, S. Vermeulen, R. Leonard and J. Keeley, *Land Grab or Development Opportunity? Agricultural Investment and International Land Deals in Africa*, IIED/FAO/IFAD, 2009.
15. FAO, *World Soil Charter*, FAO, 1982.
16. M. Held, K. Kummerer and K. Odendahl, Preserving Soils for Life, The Tutzing Project "Time Ecology", in *Proposal for a Convention on Sustainable Use of Soils*, 1998.
17. H. Hurni, M. Giger and K. Meyer, *Soils on the Global Agenda. Developing International Mechanisms for Sustainable Land Management*, Centre for Development and Environment, 2006, pp. 64.
18. P. A. Sanchez, S. Ahamed, F. Carré, A. E. Hartemink, J. Hempel, J. Huising, P. Lagacherie, A. B. McBratney, N. J. McKenzie, M. de Lourdes Mendonça-Santos, B. Minasny, L. Montanarella, P. Okoth, C. A. Palm, J. D. Sachs, K. D. Shepherd, T. -G. Vågen, B. Vanlauwe, M. G. Walsh, L. A. Winowiecki and G. -L. Zhang, *Science* 2009, **325**, 680–681.
19. R. J. Thomas, M. Akhtar-Schuster, L. C. Stringer, M. J. Marques, R. Escadafal, E. Abraham and G. Enne, *Environ. Sci. Policy*, 2012, **16**, 122–135.
20. UNEP, *Yearbook 2012*, 2012.
21. S. Jeffery, C. Gardi, A. Jones, L. Montanarella, L. Marmo, L. Miko, K. Ritz, G. Peres, J. Römbke and W. H. van der Putten, *European Atlas of Soil Biodiversity*, 2010.
22. UNCCD, *A Sustainable Development Goal for Rio+20: Zero Net Land Degradation*, 2012.
23. G. Tóth, V. Stolbovoy and L. Montanarella, *EUR 22721 EN*, Office for Official Publications of the European Communities, Luxembourg, 2007, pp. 40.

Soil Natural Capital and Ecosystem Service Delivery in a World of Global Soil Change

DAVID A. ROBINSON*[a], BRIDGET A. EMMETT[a], BRIAN REYNOLDS[a], ED C. ROWE[a], DAVE SPURGEON[b], AIDAN M. KEITH[c], INMA LEBRON[a] AND NEAL HOCKLEY[d]

[a] NERC Centre for Ecology and Hydrology, Environment Centre Wales, Deiniol Road, Bangor, LL57 2UW, UK; [b] NERC Centre for Ecology and Hydrology, Wallingford, Oxon, UK; [c] NERC Centre for Ecology & Hydrology, Lancaster Environment Centre, Library Avenue, Bailrigg, Lancaster, UK; [d] School of Environment, Natural Resources and Geography, Bangor University, Bangor, UK

*E-mail: darearthscience@yahoo.com

ABSTRACT

Soils provide important functions for society that include not only the provisioning of food, feed and fibre, but also the regulation of climate through carbon storage, the recycling of waste, the filtering of water and nutrient cycling, as well as forming a habitat for genetic diversity. These ecosystem services are supported by the soil's natural capital stocks and are important for the functioning of the earth system. Soil must therefore be managed, not with a single function in mind such as food production, but as a multifunctional resource. Ecosystem service

Issues in Environmental Science and Technology No. 35
Soils and Food Security
Edited by R.E. Hester and R.M. Harrison

Published by the Royal Society of Chemistry, www.rsc.org

concepts promote this idea, so that we can understand management trade-offs for decision making in policy. However, in order to utilise this framework we need to build the ecosystem service concepts for soils, understand the influence of man on soil stocks and services through anthropogenic soil change, and how best to monitor this soil change. This forms the basis of this chapter which reviews soil ecosystem service concepts, direct and indirect drivers of global soil change, and methods of monitoring national soil change using selected stocks from the Countryside Survey of Great Britain as an example. Finally we focus on valuation, and consider some of the methods being used to value ecosystem services and how these might be applied to the soil resource.

1 Overview of Soil Ecosystem Services

Society exploits nature to produce goods and services that are of benefit to our individual and societal well-being, food being a primary example of nature's provision. Supporting this is the thin layer of soil that envelopes the earth, lying between us, our prosperity and certain starvation. Stewarding the land and using our soil resource with wisdom and care is, therefore, of the utmost importance to our continued well-being and a sustainable society. Food, feed and fibre production, however, represent only one set of services we obtain from soils. Other, less well recognised services such as waste disposal, nutrient recycling, water filtration, carbon (C) storage and the support of genetic diversity are all vital to maintaining the functioning of the earth system as increasingly soils need to fulfil a multifunctional role.

1.1 Nature's Services

Soils are a vital component of the earth system, not only acting as the biogeochemical engine, but also fulfilling a range of important functions that include supporting and sustaining our terrestrial ecosystems, regulating the atmosphere, filtering water, and recycling waste. Increasingly, soils offer an important cultural resource, preserving artefacts and heritage, supporting landscapes, providing aesthetic beauty in the form of soilscapes, as well as recreation areas and sports fields. The faunal biodiversity of soils, long recognised, has provided important medical resources such as antibiotics,[1] and continues to provide new discoveries like the recent finding reported by the BBC[2] that the *Clostridium sporogenes* bacterium may provide a promising way of delivering cancer drugs into tumours. The challenge for society is to determine how to balance use and exploitation of the soil resource in a way that maintains all these functions. We must address the issue of how we value our soils and trade-offs in their functionality in an increasingly anthropocentric world where decisions are often made according to cost–benefit analysis.

Mankind has understood for millennia that our future well-being is intimately linked to nature on the planet which we inhabit. As far back as early biblical times, Moses – sending out spies to assess the land of Canaan – asked "How is the soil? Is it fertile or poor? Are there trees in it or not?" Moses, as the leader of a nation, understood that their fate lay with the services that nature could provide to sustain their society. In 2005 the United Nations Environment Program commissioned the Millennium Ecosystem Assessment,[3] a contemporary equivalent for evaluating our resources; the scale has increased since biblical times but the concept remains the same. This assessment made a huge global impact at a political level by bringing to light the state of the earth's ecosystems. The report identified that many ecosystems were in serious decline, or at the point of collapse. Stark reading has influenced many Governments towards adopting a 'green' approach to policy development, endeavouring to account for the goods and services we obtain from nature, to protect and enhance these by making decisions that incorporate nature's value. These overall concepts are embodied in what is termed the 'ecosystems approach' which forms the context of this chapter, especially with regard to how soil science can contribute to this.

1.2 Review of the Ecosystems Approach

In 1977 it was suggested that society could make more informed decisions and policy by incorporating the idea that ecosystems offered benefits of social value.[4] This idea has grown into the concept of what we term 'natural capital and ecosystem services'. The ecosystems approach recognises that we live in a coupled human–earth system and that many policy decisions are governed by socio-economics. As many decisions regarding the environment are made within this socio-economic framework, in order to balance the needs of maintaining a functional earth system for wider society with more local individual needs, we must integrate the value of earth's natural capital and ecosystem services into the decision-making process. The ecosystems approach, as interpreted in this work, is focused on valuing interventions required for managing land and ecosystems, and determining trade-offs for making decisions. Here we do not consider what might be termed 'intrinsic value', as might be ascribed to rare or endangered species; there are alternative, moral and ethical reasons for the protection of endangered species and other decision-making processes may be more appropriate. Therefore, the focus of the ecosystems approach is on placing value on the benefits we receive from natural products and processes, rather than engineered solutions, and also recognising the wider benefits of these products and processes, as nature is often overlooked in the decision-making process. If we wish to remain a sustainable and viable society then we must account for the utilisation of nature and ensure that societal economic growth does not depreciate our natural resources to the point of collapse, but manage our resources by recycling and replenishing them appropriately. These concepts and principles

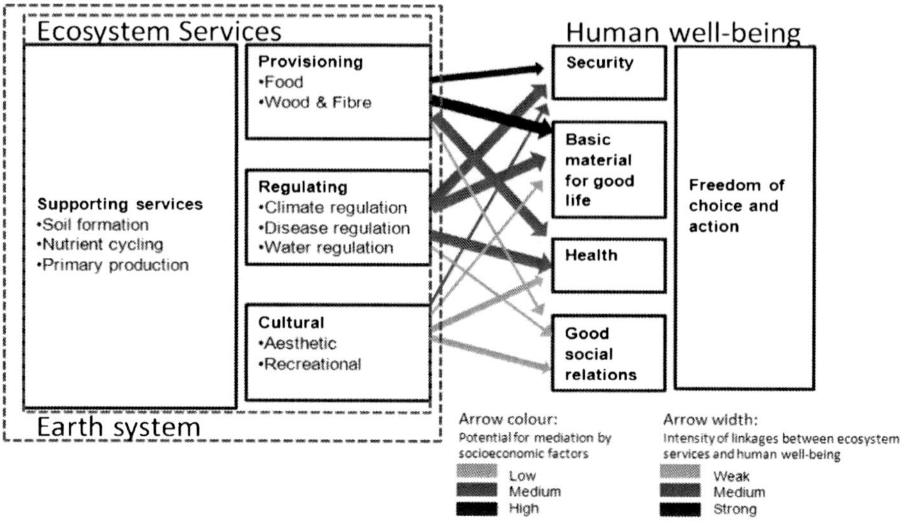

Figure 1 Diagram showing the strength of linkages between ecosystem service categories and the components of human well-being. (Source: adapted from the Millenium Ecosystem Assessment).[3] Note that soils only appear as a supporting service.

can be found in a growing body of literature including works such as *Nature's Services*[5] and *An Introduction to Ecological Economics*.[6]

The Millennium Ecosystem Assessment (MEA) has had huge impact and has been successful in bridging the science–policy divide, linking ecosystem services to human well-being and decision making. The concept presented in the 2005 report (Figure 1), with provisioning, regulating and cultural services maintained by supporting services and mapped to human well-being, has created a profound overarching framework for this. The challenge since its publication has been to take these ideas and make them into an operational system for decision making. This presents a test on many fronts; for example: what constitutes an ecosystem service; which should be valued; how should they be valued; who derives benefits? We therefore take some time here to look at how the concepts and definitions within the ecosystems approach are developing.

1.2.1 Ecosystem Goods and Services

Looking at fundamentals, a 'good' is defined as "a physical or tangible item, a product that can be seen, tasted, felt, heard, or smelled". It can be owned, and satisfies some human want or need, or something which people find useful or desirable. Conversely, a 'service' is by definition more abstract, "a type of economic activity that is intangible and insubstantial, it cannot be touched, gripped, handled, looked at, smelled, tasted or heard. It is not stored and does not result in ownership; a service is consumed at the point of sale." Thus, in terms of the commodities we obtain from nature, the goods would be food, fibre, or wood, whereas services could be carbon or water storage, cleaning, waste disposal or recycling. With regard to benefits,[7] the MEA broadly considered services and benefits to be the same. However, a benefit is not a service.[8] A 'benefit' is, in general, defined as an 'advantage', more specifically in the marketing context it is "a desirable attribute of a good or service, which a customer perceives he or she will get from purchasing." Fisher and Turner[9] point out that treating benefits and services as the same also creates a potential pitfall of double counting in valuation.

Fisher *et al.*[10] provide a recent overview of how ecosystem services have been defined in the literature, indicating that the literature has no commonly accepted consistent definition, something that they, and others,[7,8] argue is required to turn a conceptual framework into an operational system of accounting. The three following definitions of ecosystem services are most commonly cited:[10]

- The conditions and processes through which natural ecosystems, and the species that make them up, sustain and fulfil human life;[5]
- The benefits human populations derive, directly or indirectly, from ecosystem functions;[11] and
- The benefits people obtain from ecosystems.[3]

They go on to state "These definitions suggest that while there is broad agreement on the general idea of ecosystem services, important differences can be highlighted. In Daily[5] ecosystem services are the 'conditions and processes,' as well as the 'actual life-support functions.' In Costanza *et al.*[11] ecosystem services represent the goods and services derived from the functions and utilised by humanity. In the MEA, services are benefits writ large." As a result, there is often confusion as to what is meant when the term is used, the mixing of benefits and processes with services also adds to this.

In response to this confusion of terms, attempts have been made to clarify the definition of services to develop an operational green-accounting definition:[8]

> "Final ecosystem services are components of nature, directly enjoyed, consumed or used to yield human well-being".

Furthermore, they argue that an ecosystem index must be developed along similar lines to the existing labour and capital indices contributing to Gross Domestic Product (GDP). To avoid 'double counting' these indexes only count final products that are consumed by humans; for instance, cars and trucks are counted, but the component parts, such as, tyres, glass, headlamps, radios are not. This leads to distinguishing between final (car) and intermediate (parts) goods and services.[8] The recent National Ecosystem Assessment[12] for the United Kingdom has followed this concept of identifying intermediate and final services, and focusing valuation on the benefits we obtain from final services.

1.2.2 Natural Capital

Concurrent with the development of the ecosystem service concepts, other workers have placed emphasis on the term 'natural capital';[6] the term being coined by Schumacher,[13] though reference to soil and land as natural capital can be found dating as far back as 1836.[14] Costanza and Daly[15] broadly define natural capital as "a stock that yields a flow of valuable goods or services into the future". In more recent work, natural capital is defined as "the stock of materials or information contained within an ecosystem".[11] Natural capital emphasises nature's stocks, whilst ecosystem services emphasise nature's flows. Given that both are important, any overarching ecosystems framework should consider both, as sustainability aims to optimise the benefits from flows, without degrading stocks.

1.3 Soil Natural Capital and Ecosystem Services

The first attempt to classify the ecosystem services of soils was perhaps that of Daily *et al.*[16] (Table 1), which has been followed by other classifications,[17,18] especially a number for the purposes of agriculture.[19] Most of the work presented on soil ecosystem services has focused on identifying the types of services that soils deliver, especially those from the soil biota which support many soil functions. However, very little work has been done on the development or refinement of the conceptual framework for soil ecosystem service delivery. This presents an important problem, since under the MEA classification soils provide supporting services (Figure 1), and therefore do not appear in any final valuation.

In parallel to these efforts, soil scientists have identified with the ideas of natural capital, which is perhaps more intuitive to soil science dealing with stocks. Initial attempts to make the term more concrete defined the natural

Table 1 Societal soil ecosystem services. (Source: adapted from Parkinson and Costanza).[2,15]

Societal soil ecosystem services
SUPPORTING
Physical stability and support for plants
Renewal, retention and delivery of nutrients for plants
Habitat and gene pool
REGULATING
Regulation of major elemental cycles
Buffering, filtering and moderation of the hydrological cycle
Disposal of wastes and dead organic matter
PROVISIONING
Building material
CULTURAL
Heritage sites, archaeological preserver of artefacts
Spiritual value, religious sites and burial grounds

capital of soils as texture, mineralogy and soil organic matter.[20] A more comprehensive assessment of soil natural capital defined it as "the stocks of matter, energy and organisation within soils".[21] This classification breaks down soils into their fundamental components and considers the quality and quantity of each component where applicable (Table 2). The assessment of stocks in the natural capital approach is important for soils because flows can be determined from a change in stock, but stocks must be assessed at least once to determine how they will change from flows, and sometimes we need to know stock to determine sustainability. Secondly, soil science has a vast collection of soil resource data, but most of this is in the form of stock assessment, so in order to utilise this information in ecosystem service mapping it is advantageous to consider stock and change.

Dominati *et al.*[22] were the first to attempt to unify the concepts of ecosystem services and natural capital for soils. They recognised the importance of combining both frameworks, and seeing them as mutually compatible in expressing an ecosystem approach for soils. They provide a comprehensive overview of progress in both natural capital and ecosystem services for soils. What the work[22] recognised above all was this need for a unified approach to develop an operational framework for soils.

In parallel, but not unconnected work, the issue of investing in our 'ecological infrastructure' was raised;[23] it was argued that "there is also an ecological infrastructure that maintains the provision of the ecosystem services that support a wide range of ecological as well as socio-economic benefits." This concept is an emerging one, but particularly suited to soils, which do provide many support roles and intermediate services within ecosystems and often require investment in the natural capital soil stocks in order to maintain their full range of function.

Table 2 Soil natural capital in the matter, energy, organisation classification.[21]

	MEASURABLE OR QUANTIFIABLE SOIL STOCK
1) MATTER	
Solid	Inorganic material (I) Mineral stock & (II) Nutrient stock Organic material (I) OM/Carbon stock & (II) Organisms
Liquid	Soil water content
Gas	Soil air
2) ENERGY	
Thermal Energy	Soil temperature
Biomass Energy	Soil biomass
3) ORGANISATION (ENTROPY & INFORMATION)	
Physico-chemical Structure	Soil physico-chemical organisation, soil structure
Biotic Structure	Biological population organisation, food webs and biodiversity
Spatio-temporal Structure	Connectivity, patches and gradients

In summary, with interest growing and concepts emerging and developing, the challenges for the soil science community to develop an operational soils component for the ecosystems approach have been stated as:[24]

- Creating the appropriate frameworks to determine the natural capital and intermediate-and-final goods and services supplied by soils that benefit human well-being, maintain the Earth's life support systems, and promote biodiversity.
- Identifying appropriate measurement and monitoring programmes with agreed metrics to develop the evidence base on the 'state and change' of soil natural capital and the ecosystem services that flow from it.
- Developing the means to value benefits from soils which can feed into the frameworks being developed in other disciplines and, where possible, develop synergy with existing national accounting frameworks such as GDP and state-of-the-environment (SoE) reporting.
- Engaging in the development of decision support tools that incorporate 'soil change', that will enable the most informed comparison of trade-offs in the decision-making process, cognisant of the enormous practical challenges this implies.

In the following section we explore a synthesis, developing a stock-and-flow framework that, like Dominati's,[22] brings together the ecosystem service and natural capital concepts.

1.4 Stock-and-flow Ecosystem Service Framework for Soils

A valuation index, similar to GDP, helping us to make our economies more sustainable is one major goal of the ecosystems approach. GDP focuses on the value of final goods and services, the major criticism being that there is no accounting for resource depletion and degradation. Ecosystem service concepts must avoid this pitfall if they are to give a reasonable representation of ecosystem service delivery as well as resource use to aid sustainable delivery. Therefore, we need some system that tracks not only the ecosystem services delivered, but the state of the soil resource as services are delivered. This encourages us to think of ecosystem service delivery in terms of a supply chain. Soils are composed of fundamental stocks of 'matter, energy and organisation' – these are the building blocks. Processes act on these stocks, combining and transforming them, resulting in intermediate goods and services from which society doesn't derive direct benefit; these stocks and intermediate goods and services form the soil ecological infrastructure.[23] If we are to achieve sustainability we must understand the linkage between the health of the ecological infrastructure and the delivery of final ecosystem goods and services. Sustainable options are often more expensive than short-term resource mining because effort is expended in maintaining the infrastructure.

The stock-and-flow framework then recognises that soils may contribute final ecosystem services by following one of two routes, which can be thought

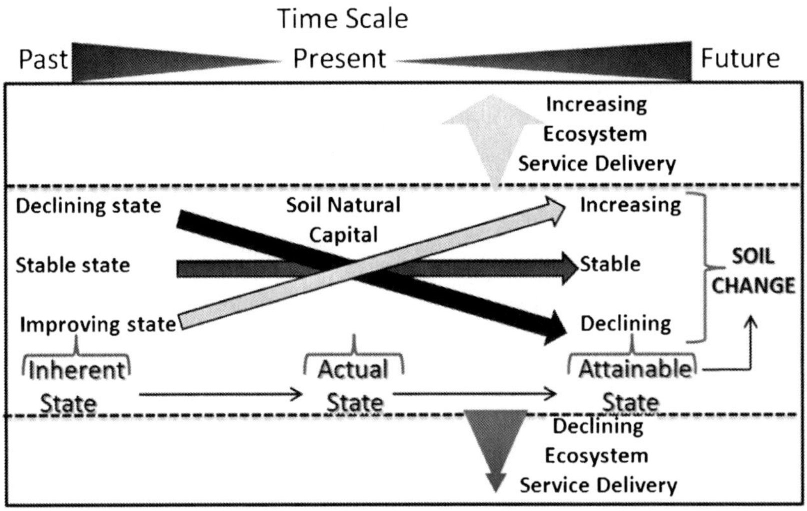

Figure 2 The temporal balance between soil natural capital (the central thick arrows) and ecosystem service delivery as a function of soil change. (Source: adapted from Bristow).[23] The actual state describes current stocks, whereas the inherent state describes that which the soil comes from, and the attainable state is that which may be obtained through management in future. A declining state indicates soil resource depletion, whereas an increasing state indicates soil resource accumulation. If soil resources decline, ecosystem service delivery is also likely to decline.

of as explicit and emergent. The explicit route includes soil goods and services that we consume directly, *e.g.* topsoil, peat, *etc.*, whereas the emergent route is where soils are only one component that contributes to the delivery of final goods and services from which we derive benefit, *e.g.* clean water delivery or flood alleviation.

If this approach is to be used then it is vital that we understand how soils are changing, both through natural and anthropogenic soil change. Figure 2 shows the link between soil natural capital and ecosystem services and the concept of soil change, which provides the time element over which change occurs. Decision making will be concerned with what the current state is and what the attainable state will be, faced with a number of land-management options. The impact of decisions can then be monitored; the Countryside Survey in the United Kingdom[25] is one example of this, and is discussed later. By valuing the changes in stocks, assessment can be made of the impact of policies and decisions made. Clearly, this approach requires a firm understanding of soil change, which forms the focus of the next section.

2 Drivers of Global Soil Change

Pedology, over the last 100 years, has focused extensively on the gradual change resulting in soil formation, encapsulated in Jenny's five factors of soil formation:[26] CL, O, R, P, T (CLimate, Organisms, Relief, Parent-material, all as a function of Time). However, there is growing recognition that anthropogenic activity has a strong influence on soil development and formation, which is more commonly being referred to as anthropogenic soil change.[27] Estimates for the next 50 years indicate that mankind is moving to a global density of 1 person for each 0.01 km^2 of reasonably biologically productive land.[28] This increase in population pressure means that we must continue to extract more from our soils to support the growing demand. This section focuses on how mankind's interaction with the land, both directly and indirectly, results in soil change, the consequent alteration of soil stocks-and-flows, and the delivery of ecosystem services on time-scales of relevance to policy making.

2.1 Direct Drivers of Soil Change

Tillage, traffic, irrigation, fertilisation and pesticide application are perhaps the main drivers of direct anthropogenic soil change and are mostly related to food production. Of the Earth's terrestrial land surface, *ca.* 134 million km^2 (Mkm^2), arable agriculture was estimated to cover 15 Mkm^2, and managed grazing[29] 28 Mkm^2, whilst the amount of land irrigated was estimated to be 2.7 Mkm^2 in 2000.[30] This means that *ca.* 38% of the Earth's ice-free land surface is currently used for agriculture, with this land replacing forests, savannas and grasslands.[31]

2.1.1 Physical and Biogeochemical Soil Change as a Result of Food Production

Tillage changes a range of soil properties including structure, density, porosity, moisture, aeration, greenhouse gas emission[32] and temperature; chemical properties, *e.g.* carbon content; and biological structure, *e.g.* earthworm population and fungal/bacterial ratio.[33] Figure 3, reprinted from ref. 31, shows the impact of tillage on soil organism groups, indicating that the larger soil organisms, such as worms, spiders and beetles, are much more susceptible to tillage activities. A recent meta-analysis of data indicated that when land use changed from forest to crop it resulted in a 42% reduction in soil C stock, whilst from pasture to crop was a 59% reduction.[34] Not only does this mean that this C is released into the atmosphere, but it also means that soil structure declines, as does the ability of the soil to retain nutrient and water stocks. Moreover, soil organic matter provides energy for the soil biota which are primarily responsible for the biogeochemical cycling in soils and the delivery of important ecosystem regulating services such as filtration and waste disposal. Traffic, both by vehicles and animals, causes soil compaction which increases soil strength and decreases soil physical fertility through decreasing storage and supply of water and nutrients. The global extent of vehicular soil compaction is estimated to affect 0.68 Mkm^2 of land;[35] that due to animals is

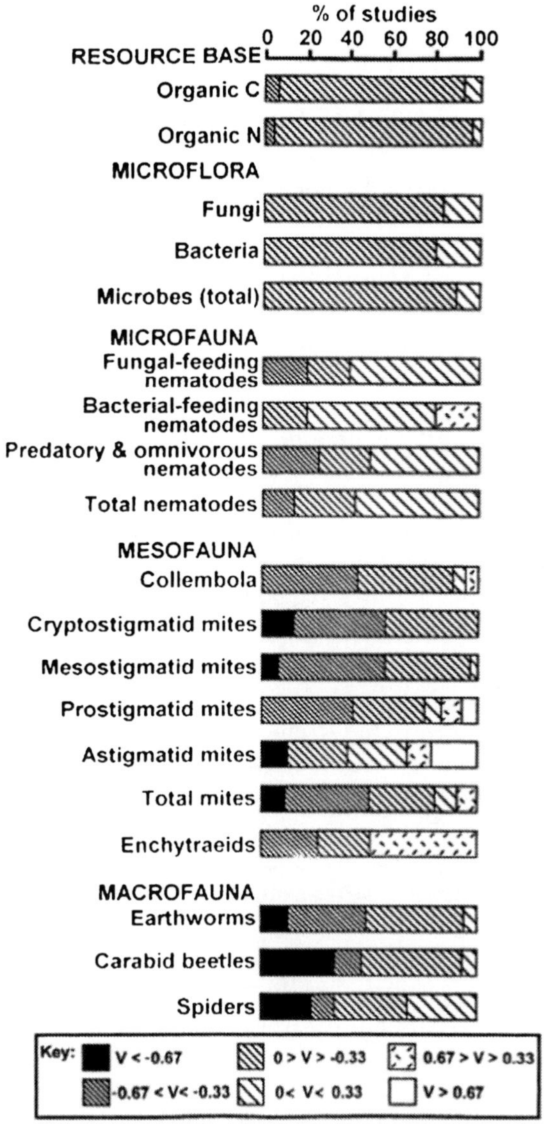

Figure 3 Results of studies compiled in ref. 75, in which the index V represents the relative difference in abundance or biomass of organism groups between no-till and conventional tillage. The index V ranges from −1 to +1 and is increasingly negative or positive as the group is increasingly harmed or enhanced, respectively, by tillage operations. (Reprinted with kind permission from Foley *et al.*).[31]

not known. This combination of tillage and compaction contributes to enhanced soil loss from erosion. It is estimated that 10.94 Mkm^2 of land is affected by water erosion, of which 7.51 Mkm^2 is severely affected, and 5.49 Mkm^2 by wind erosion, of which 2.96 Mkm^2 is severely affected.[36]

Irrigation is required to turn marginal dry lands into productive agricultural land. Soil moisture not only provides water for plant growth, but is a vital soil stock whose quality (pH, salinity) and quantity (volumetric water content) control many soil processes.[37] Life depends on water, and soil is no different, with the soil moisture content controlling the soil microbial activity and, in turn, the soil biogeochemical cycling. However, if irrigation is managed poorly it results in waterlogging and salinisation. In the San Joaquin Valley of California, modelling studies have shown how irrigation practice over the last 40 years has led to increased soil salinisation.[38] It is estimated that approximately 0.45 Mkm^2, representing 20% of the world's total irrigated land, suffers from salinisation or waterlogging.

Addition of chemicals to soils (in the form of both fertilisers and pesticides to enhance productivity) brings about soil change. According to Foley *et al.*[39] there has been a *ca.* 700% increase in global fertiliser use during the past 40 years. In the case of reactive N, production *via* the Haber–Bosch process has gone from 0 prior to 1910, to more than 100 Tg N in 2000, with 85% going into fertiliser production.[40] Fertiliser application has helped increase food production, but addition tends to acidify soils and creation rates are faster than denitrification, so reactive N is building up in the environment. It has been proposed that reactive N levels have already exceeded planetary boundaries,[41] where the boundary represents a safe operating space for humanity. Fertiliser application has a range of effects on soils; obviously it increases nutrient content, and food production wouldn't be sustainable without them. The long-term Broadbalk (Rothamsted) experiment that has grown continuous wheat since 1843, shows a strong linear correlation between the amount of fertiliser N applied annually (ranging from 0–200 kg N ha^{-1}) and the quantity of organic C accumulated in the soil.[42] Long-term fertiliser applications have also been reported, in a number of cases, to cause increases in water-stable aggregation, porosity, infiltration capacity and hydraulic conductivity, and decreases in bulk density as compared with unfertilised crop production. However, detrimental effects are observed when large doses of ammonium (NH_4^+) salts are added to soils, causing colloid dispersion in fine, poorly aggregated soils, and – if persistent – can cause crusting and reduced infiltration.[42]

Pesticides include a wide range of organic and inorganic materials, applied to kill different organisms such as weed plants and disease-causing fungi and nematodes. These materials have a direct effect on the target organisms, but may also have indirect effects on the structure and function of wider biotic communities in soils. Chemicals can cause alteration of metabolism of endemic soil microorganisms and arthropods, or result in the eradication of some components of the primary food chain. Although early research was upbeat that pesticides had no major detrimental impact on soil biota, more recent work indicates that pesticides, and particularly repeated application, can result in a range of effects on the soil biota.[43]

2.1.2 Artificial Soil Sealing by Man-made Infrastructure

Artificial soil sealing is the direct loss of soil by its covering with urban infrastructure in the form of roads, buildings and recreation areas, *etc.* This generally results in a loss of function, though some storage functions may be maintained. The amount of soil sealed globally is not known, but urban land cover, as of 2000,[44] has been estimated to be 0.4 Mkm^2. With increasing urbanisation, it is projected that urban infrastructure will cover a land area of *ca.* 1 Mkm^2 by 2030. Estimates of sealing in Europe suggest that 9% of European land area is now covered by some form of impermeable infrastructure.[45] Sealed soils are changed because they no longer interact freely with the other compartments of the Earth system. Thus the movement of matter and energy is severely inhibited, as is the ability to deliver ecosystem services, other than perhaps the protection of heritage as a cultural service.

2.2 Indirect Drivers of Soil Change

Mankind is transforming the Earth system through climate and land-use change as well as the movement of invasive species. Soils, forming the thin interface at the Earth's surface, interact with the atmosphere, hydrosphere, lithosphere and biosphere, so that as they alter, soils are often also altered.

2.2.1 Changes in the Atmosphere

The composition of the atmosphere changes, due to both natural phenomena and man's activities. The industrial era has been marked not only by increasing atmospheric CO_2 concentrations, but also by more SO_2, reactive N species, organics, heavy metals and dust in the atmosphere than in pre-industrial times.[46] Global emissions of NO, NH_3 and SO_2 are estimated to have increased by more than a factor of 3 since the pre-industrial era, largely due to the use of fossil fuels and agricultural production.[47] Modelling global deposition estimates that 36–51% of all NO_y, NH_x, and SO_x are deposited over the ocean and that 50–80% of that deposited on land falls on non-agricultural vegetation. The 'critical load' threshold, if taken to be *ca.* 1000 mg(N) m^{-2} yr^{-1}, is considered to be exceeded on 11% of the world's non-agricultural vegetation. Two major consequences for soil change are increased acidification and enhanced ecosystem productivity, with implications for the carbon cycle and biodiversity.[47]

Soils also accumulate materials from aerosol deposition occurring from both marine and terrestrial sources as mineral aerosols (*e.g.* dust).[48] The major ions contained in marine aerosols are Na^+, K^+, Mg^{2+}, Ca^{2+}, SO_4^{2-} and Cl^-, and the addition of these to soils through precipitation can be important, especially in nutrient-depleted soils near coastal regions, but also in soils of the continental interior. Mineral aerosols occur due to wind erosion of soils and sediments. At one time, agricultural soils were considered to contribute as much as 50% of the annual global dust load. However, recent work suggests a more realistic figure is that dust from agricultural areas contributes <10% to the global dust load;[49] this means that soils are receiving dust inputs from either sediments or soils, ostensibly in dry lands, especially playa landforms. We know that dust input to soil can be substantial with the formation of large areas of loess soils in China and the NW USA. The annual global dust flux is estimated at 1700 Mt yr^{-1} based on an average of five studies,[50] with *ca.* 25% estimated to deposit in the oceans. This means *ca.* 1275 Mt yr^{-1} ends up on the land, with a large proportion of this likely being deposited on soils.

A range of other materials are deposited on soils from the atmosphere, including organic and inorganic pollutants. Soil is considered to play an important role in retarding the global recycling of persistent organic pollutants from industrialised northern countries, whilst soil also accumulates heavy metals increasingly deposited from the atmosphere.

2.2.2 Changes in the Hydrosphere

Alteration of rainfall patterns is chiefly associated with climate change due to natural perturbation and human activity.[51] This can result in regional changes in the amount of precipitation, the timing or frequency of events, or the intensity of precipitation. Complex impacts are observed; for instance, increased rainfall in the Chihuahuan Desert has led to increased soil moisture and increase in woody vegetation.[52] Changes in soil moisture pool dynamics are not only important for plant growth and biogeochemical cycling,[37] but also for buffering surface temperatures. For instance, the recent increased intensity and persistence of heatwaves over Europe has been attributed to a reduced soil moisture buffer.[53]

Changes in the levels of groundwater have also been observed to cause detrimental soil change. In Australia, for example, the removal of trees in the Murray Darling Basin caused a rise in the level of groundwater rich in soluble salts. As a result, large areas of soils in the basin have become salinised.[54]

2.2.3 Changes to the Lithosphere

It is not changes to the lithosphere *per se* that lead to significant soil change, but the waste materials that are a by-product of mineral exploration. Terrestrial exploration for oil and gas can result in re-used waters high in soluble salts, metals and harmful organic materials. Mining activities produce spoil that, over time, can infuse high levels of metals and trace elements into surrounding soils. Hard-rock quarrying is less likely to lead to the release of toxins, but results in major disturbance of soil from quarried and spoil areas.

2.2.4 Changes to the Biosphere

Perhaps the largest pressure on the biosphere has been land use change, the planting and removal of trees, and the transition to monocultures in agriculture and plantation forestry. Vegetation has a fundamental impact on soil formation and behaviour. Meta-analysis of afforestation research, principally in temperate latitudes, showed that the extensive planting of forest monocultures has been associated with depletion of soil Ca^{2+}, K^{+} and Mg^{2+}, increase in Na^{+}, and a mean pH decrease of 0.3. In addition, soil C and N were observed to decrease, though principally under pine.[55] Conversely, recent research from the tropics has indicated that in very acid tropical soils, planting with pine may actually increase soil pH.[56] A further change observed with forestry, especially pine plantations, is the development of soil water repellency, both through afforestation and exacerbation by forest fire.[57] Soil water repellency can alter soil hydrological behaviour and enhance soil erosion.

Biosphere change indirectly affects soil biota and the processes that they drive, by altering plant community composition, nutrient cycling, carbon allocation patterns, or the quantity and quality of plant-derived organic materials.[58] Many of these processes, and linkages between above and below ground, are not well understood. Many researchers suggest a high degree of functional equivalence in the soil decomposer community, and hence substantial redundancy in species richness and diversity, which makes it difficult to assess the impacts of global change on soil biota.

Pollution of the biosphere, for instance by heavy metals, can cause harm to soil ecosystems. A number of keystone soil taxa, *e.g.* earthworms, springtails, and nitrifying bacteria, are particularly sensitive to metals. High heavy-metal concentration in soils can reduce the abundance and diversity of communities of these and other taxa, potentially resulting in a breakdown of specific soil functions such as decomposition, nutrient turnover and the regulation of hydrological flows through soil and the resulting delivery of ecosystem services. Metals are not broken down over time (unlike most organic chemicals) and so can be removed only by the relatively slow process of cropping and leaching. The accumulation of metals presents one of the more serious long-term threats to soil sustainability worldwide.

2.3 Mapping Global Soil Change

The development of soil mapping stems largely from the need to identify and value land for the production of food, feed and fibre. This has resulted in inventories of soils and maps of their spatial distribution, often based on soil development characteristics such as soil horizons. With regard to the ecosystems approach, there is much more of a need to understand how soils are changing, which is not easily determined from the classical soil inventories and maps. The need to understand how soils change has led to attempts to resurvey soils, such as that for England and Wales for soil C content.[59] Moreover, dedicated monitoring frameworks have been designed to detect soil change, the longest running of which is the Countryside Survey for Great Britain,[25] which has monitored ecosystems and soil stocks since 1978. Similar monitoring programs are now run Europe-wide for agricultural soils under the LUCAS monitoring programme[60] and for forest soils under the BIOSOIL programme.[61] There is much scope for discussion as to how these stocks are determined, but the approach lays the foundation for a quantifiable assessment of soil stock and change, from which the impacts of management and ecosystem service delivery can be determined. The following section focuses on the Countryside Survey, examining soil stock and change for Great Britain.

3 National Soil Change, the Countryside Survey of Great Britain

Determining drivers of soil change at national levels is key to understanding sustainable soil ecosystem service delivery. Obtaining a national overview of all habitats is also important for discriminating between change occurring due to practices such as agriculture and change due to other natural or anthropogenic drivers. Countryside Survey (CS) is a unique study or 'audit' of the natural resources of the British countryside that aims to do this.[62] The sampling strategy for CS is based on a stratification of Great Britain (GB) into Institute of Terrestrial Ecology (ITE) Land Classes defined by the major environmental gradients across the countryside.[63] The sample is structured to give reliable national statistics and to ensure that the range of different environments found in GB is adequately represented. The sample consists of a set of 'sample squares' measuring 1 km × 1 km, selected randomly from the GB Ordnance Survey grid within the various ITE Land Classes. The Survey has been carried out at regular intervals since 1978 using consistent and rigorous scientific methods. Monitoring of soil stocks (topsoil 0–15cm), and their change, has been an important part of the program since 1978, and the number of soil stocks monitored continues to grow to address policy questions of national importance to the UK. Some of the key policy questions to be answered by the latest survey in 2007 were:

- Can the loss of soil C (0–15cm) as reported by Bellamy *et al.*[59] be confirmed?
- Has the recovery from acidification detected by CS in 1998 between 1978 and 1998 continued?
- Can the trend of eutrophication of the countryside detected in the vegetation also be detected in the soil using the mean total N concentration?
- Can the trend of eutrophication of the countryside detected in the vegetation also be detected in the soil using a more sensitive soil process method for N?
- Is the decline in atmospheric deposition of heavy metals as reported by the Heavy Metals Monitoring Network reflected in soil metal concentrations measured in the CS?
- Does the CS provide any evidence to indicate that there has been a loss of soil biodiversity as has been stated by the European Union?

By measuring soil stocks, and determining 'soil change' at the same location as other measurements of ecosystem change, the CS is uniquely capable of assessing changes in ecosystem service delivery at a national scale through a stock and flow framework, some of the major findings of which are now presented.

3.1 Carbon

The soil organic carbon (SOC) stock is one of the headline indictors of soil quality and there is a wide acceptance that C is fundamental to soil functioning as it is the primary energy source in soils and has a critical role in maintaining soil structural condition, resilience and water retention; all soils therefore need to retain C. However, soil C changes are measured against large background stocks and high spatial heterogeneity, and more information is needed to be able to manage this resource better. Specific policy requirements which require improved information on the status and change of soil C content include UK and EU legislation soil protection measures that will help to conserve soil C. The reformed Common Agriculture Policy requires all farmers in receipt of the single payment to take measures to protect their soil from erosion, organic matter decline and structural damage. Changes to soil C content also represent a major component of UK greenhouse gas emissions and under the Kyoto Protocol the UK is required to make estimates of net C emissions to the atmosphere. However, knowledge of soil C stocks and changes is limited; recent work by the National Soil Resources Institute (NSRI)[59] indicated that large changes have occurred recently, but there has been some debate concerning possible causes.[64] Based on C concentration data in the soil (0–15 cm) collected from Great Britain between 1978 and 2007, the CS was unable to support this finding, with data showing that there had been no overall change. Area estimates for each Broad Habitat were used to convert soil C densities to soil (0–15cm) C stock for GB and its component countries. Soil (0–15cm) values were 1582 Tg C for GB and 795 Tg C, 628 Tg C and 159 Tg C for England, Scotland and Wales in 2007, respectively. It must be emphasised this significantly underestimates the total C stock in soils due to the large C stores at depth, particularly in peat soils.

3.2 pH

Soil pH is perhaps the most important indicator of soil chemical condition, and the most commonly measured. It is considered one of the quality components of the soil water content in the soil natural capital classification (Table 2). It gives an indication of soil acidity and alkalinity and therefore has direct policy relevance both to recovery from soil acidification and to any soil alkalinisation that may occur due to land use and or climate change. A strong correlation between bacterial diversity across GB and soil pH has been found,[65] suggesting changes in pH may affect soil bacterial biodiversity. In addition, soil pH is important for predicting nutrient availability and the mobility and bioavailability of metals in soils. It is currently estimated that 58% of terrestrial semi-natural habitats across Great Britain receive acidic deposition in excess of their buffering capacity, thus potentially causing long-term damage according to the critical load methodology.[66]

Soil pH data from the CS between 1978 and 2007 has given a unique national assessment of soil pH change over this time period. It was found that pH increased over the time period in the majority of Broad Habitats, with soils in general becoming less acidic. This agrees with recent independent work[67] and is consistent with the expected benefit of continued reductions in sulfur emissions. Data analysis continues to explore whether the smaller sulfur deposition reductions experienced in the north (Scotland) and west (Wales) of GB or other drivers of change, such as N deposition and land management, are responsible for a lack of significant pH increase between 1998 and 2007 in organic-rich soils that occur most commonly in this part of Britain. Land use change was presumably not responsible for this pH change, as detailed analysis separating plots where Aggregate Vegetation Class (AVC) had remained unchanged from those where change had been recorded showed similar pH trends.

One of the impacts of an increase in the mean soil pH across GB is that there has been a concurrent increase in the number of soils with pH above 8.3.[68] The theoretical pH of a solution exposed to the atmosphere ($p_{CO2} = 10^{-3.5}$ atm) in the presence of calcite is 8.3. The partial pressure of CO_2 in soils can be more than 100 times higher than in the atmosphere so that pH values from 7.5 to 8.5 may indicate calcite saturation. When pH is controlled by the calcite system only, soils do not in general exhibit structural problems since the abundance of Ca provides good aggregate stability. However, when monovalent cations like sodium accumulate in the soil, the pH can rise above 8.3 causing dispersion of soil colloids, soil sealing, reduced infiltration rates and enhanced soil erosion. Countryside Survey data collected from 1978, 1998 and 2007 showed that the mean soil pH for locations with pH >8.3 and sampled in all three surveys had experienced an increase of *ca.* 1 pH unit from 1978 to 2007. The majority of these samples were taken from calcareous soil types or over calcareous parent material, and predominantly in the Arable Broad Habitat type; farming tillage practices may therefore be bringing calcareous minerals to the surface, which when affected by sodium can cause a further increase in the pH. Sources of sodium may include mineral weathering, salting of roads or atmospheric deposition. Moreover, the region where soil pH is increasing is coincident with the greatest decrease in acid atmospheric deposition over the last 20 years, enhancing the stability of alkali minerals.

The implications of these findings are that current emission control policies, combined with policies to protect soil through sustainable land management practices, have had some major benefits in reducing acidity. However, there may be a trade-off as alkaline soils may increase in abundance and become more susceptible to structural problems if sodium levels in the environment increase, *e.g.* through winter salting of roads and the blowing of salt onto calcareous arable land; targeted monitoring of the sodicity of vulnerable soils should be performed in order to avoid soil structural decline.

3.3 Nitrogen

Nitrogen (N) is the fourth-most abundant element in living organisms, but is energetically expensive to acquire and is easily lost from ecosystems through leaching and trace gas losses. For these reasons, N availability limits plant production in most terrestrial ecosystems,[69] and soil N stock is a fundamental measure of soil fertility. Total soil N stock is large in relation to annual influx through N_2 fixation or fertiliser inputs, and usually changes slowly, although events such as forest clearance or the ploughing of permanent pasture can result in large mineralisation fluxes and N losses. Provisioning services and carbon sequestration are generally increased by large N flows, although more productive systems are often less biodiverse and species loss due to eutrophication by N *via* water and air pollution is a global problem.

Much of the N in soil is bound in organic material and not immediately available to plants. The ratio of total N to total C stock is generally considered an important indicator of N availability. However, increased N inputs can increase soil total C : N ratio, since increased productivity results in greater inputs of fresh plant material which has a larger C : N ratio than older soil organic matter. This is probably the reason for an increase in C : N ratio in several UK broad habitats and an overall decrease in total N concentration in UK soils between 1998 and 2007, in top (0–15 cm) soil.[25]

Measurements of the stock of plant-available soil N are potentially more useful than total N stock for predicting effects on ecosystem service delivery. Plants were formerly considered to take up only mineral N (nitrate and ammonium ions), but in recent decades it has become appreciated that amino acids and even small polypeptides are directly taken up by plants.[70] Plants can take up N even from soils with zero net mineralisation flux by intercepting soluble N before it can be re-immobilised, and the presence of plants commonly changes mineralisation rates by increasing sink strength or by increasing the supply of labile C to decomposers. Despite these difficulties, the stock of readily mineralisable N remains a useful comparative indicator of plant-available N. Mineralisable N stock was only measured in the most recent UK Countryside Survey (2007), but likely temporal changes can be inferred by assessing stock in relation to atmospheric N deposition rate in unfertilised habitats. The effect of increased N deposition on mineralisable N stock (in 0–15cm soil) was stronger in soils with greater C concentration,[71] presumably since much of the organic matter in more-organic soils is unreactive and so does not immobilise N additions.

The mechanisms, by which total N stock contributes to a flow of N in usable forms, and the controls on the important fluxes, are fairly well understood. However, predicting and controlling N flows from soil with sufficient accuracy to synchronise them with plant uptake remains a major challenge. Farmers have become increasingly adept at targeting applications of mineral N fertiliser in arable systems, but using organic fertilisers and ensuring that N is used efficiently within relatively extensive agricultural systems are key areas of research for the 21st century.

3.4 Metals

In the European Union and internationally, a set of research programmes have focused on assessing the risks of trace metals to ecosystems. This work has been driven by policy initiatives which include new procedures for the mandatory risk assessment according to European Commission regulation 1488/94 and studies to support the 1998 Convention on Long-Range Transboundary Air Pollution Aarhus Protocol on Heavy Metals.[72] High concentrations of metals in the environment are a threat to both soil function and the delivery of ecosystem services. This is because exposure to sufficiently high concentrations of trace elements is associated with negative effects on species, including a range of ecologically important soil taxa.[73] Most problematically, unlike the majority of organic chemicals, trace metals are not subject to degradation. Instead, once deposited to land, metals can only be removed by the relatively slow processes of sediment transport and leaching (and cropping in some systems).

At the national scale, concentrations of trace metals have been relatively well investigated,[74] including previous surveys of trace metal concentrations in soil across England and Wales.[75] Within the GB-based Countryside Survey project, trace metal concentrations have been measured in soils (0–15cm) from two surveys conducted in 1998 and 2007 in order provide baseline data on topsoil trace metal concentrations in support of risk assessment. Instrumentation developments between surveys, including a transition from inductively coupled plasma optical emission spectrometry to inductively coupled plasma mass spectrometry and the adoption of microwave digestion methods, make direct comparison between surveys difficult. These issues can, however, be overcome by the comparative analysis of certified reference materials and individual samples – with these data then used to "normalise" differently analysed data-sets.

Comparisons between the two survey data-sets available for the Countryside Survey project indicated that only small relative changes in soil trace metal concentrations occurred between surveys. This is despite reported declines in atmospheric deposition and is presumably due to the long residence time of metals in soils. Of seven metals for which repeat measurements were made during the 2007 survey, for only one, Cu, was a statistically significant difference in soil (0–15cm) concentrations (an increase) found for Great Britain. When the data for repeat metal measurements were stratified by Broad Habitat, AVC, and soil organic matter category, further statistically significant differences were seen. However, because of the need to normalise data-sets between surveys, these potential differences require further investigation to test whether they are real or experimental artefacts.

For Cu the differences seen were generally characterised as significant increases. These were consistent across a range of habitats and soil types, supporting the validity of the observation of change. For two of the metals, Cd and Pb, changes were small and idiosyncratic between stratifications, suggesting that the levels of these metals in soil are stable and that any continued deposition is low or offset by similar loses. For three metals, Cr, Ni, and Zn, changes that were seen were generally characterised by reduction in crop lands and no change, or slight increases, in less managed habitats.

Although direct comparisons between the two different surveys are challenging, it is possible to draw some conclusions that may be useful to focus further investigations. For example, for some metals, such as Cu, there is the suggestion that concentrations in soils may still be increasing. Sources of this metal, such as animal manures and possibly sewage sludge, as well as aerial deposition, may be important in maintaining or even increasing soil concentrations. For the remaining metals, there is no clear and consistent evidence of a general increase in soil concentration. Instead, there is some suggestion that in areas where cropping takes place output fluxes may now exceed inputs, enabling soil levels of some element, such as especially Cr, Ni and Zn, to decline. Consequently, managed landscapes, where intensive cropping takes place but where sewage sludge, animal manures and composts are rarely applied, may be among the first habitats to show the signs of a return from their moderately elevated states to their pre-industrial background concentrations.

3.5 Biodiversity

The activities of the soil biota are critical for the provision of many important soil functions and resulting ecosystem services. These functions include, but are not limited to, biomass production, storing, filtering and transforming nutrients, contaminants and water, and acting as a biodiversity pool from which future novel applications and products can be derived. Because they are intimately involved in many important soil functions and are fundamental to maintaining soil quality, the biological components of soils have considerable potential as indicators of soil quality. At present, comparatively little is known about the biodiversity of soil compared to, for example, above-ground diversity. In the microbial realm, molecular approaches such as terminal restriction fragment length polymorphism (tRFLP), phylogenetic microarrays and molecular fingerprinting are starting to reveal patterns in the diversity and distribution of soil microbes[65]. For the soil meio-, meso- and macro-fauna, molecular approaches to biodiversity assessment are still in development. For some key groups, such as earthworms and springtails, good keys to the UK and other national fauna do exist, although, even in these cases, traditional morphological taxonomy can be hampered by cryptic speciation and other taxonomic uncertainties.

Beyond these better-known groups, in taxa like mites and nematodes there are significant issues for morphological identification associated with the lack of keys, laborious nature of the work and declining expert base. Because they are often poorly known, few soil species are recognised for their conservation value. A few, such as some ants and fungi, are covered by the UK Biodiversity Action Plan[76] (UK BAP). However, as knowledge of soil biodiversity is often sparse, it is not well known if and how climate, land use, land management change and pollution affect populations of even these relatively well known and highly valued species, let alone the large amount of often hidden diversity that maybe present in soils. Beyond the few soil species covered by BAPs, there is a well-recognised need to quantify soil biodiversity and determine whether it is possible to observe consistent patterns of population and community structure against a dynamic background of spatial and seasonal variability. Key questions relate to determining if and how soil biodiversity changes over time, and the nature of the environmental drivers of any such changes.

National-scale surveys of soil biodiversity are relatively rare in the published literature[77]. Soil invertebrates have been identified and counted in two Countryside Survey project in 1998 and 2007, with the focus on measuring the abundance and broad taxa richness of the soil mesofauna. In the most recent survey in 2007, there were an estimated 12.8 quadrillion (1.28×10^{16}) soil invertebrates present in the top 8 cm of the soils of Great Britain during the time of sampling. Comparing these results with those from the survey in 1998 enabled change in soil biodiversity to be estimated at a national scale. A significant increase in total invertebrate catch in samples from soils (0–8cm) from all Broad Habitats, Aggregate Vegetation Classes and soil organic matter categories, except for agricultural areas on mineral soils, was found in the Countryside Survey in 2007. The increase in invertebrate catch was mainly the result of an increase in the catch of mites in 2007 samples. This resulted in an increase in the mite: springtail ratio, but a decreased Shannon diversity due to the dominance of mites in 2007 cores. A small reduction in the number of soil invertebrate broad taxa was found, which could suggest that there may be a declining trend in soil biodiversity. However, repeat sampling is required to ensure that seasonal conditions in the two sampling years, including land management, annual weather patterns or merely natural population variation, do not explain the observed changes before any general trends can be validated. Individual site studies with time-series data would be a useful place to start in this regard.[78]

Large-scale changes in soil biodiversity would undoubtedly have important consequences for ecosystem services delivered by the soil. However, these services provided by soil organisms are generally undervalued in agricultural systems since it is human inputs rather than natural processes that are considered to drive the system. For example, it was estimated that earthworms add 723 million euros per year to livestock production in Ireland.[79] Clearly, agricultural practices which promote soil's natural capital and self-regulation are needed to realise sustainable production.

4 Approaches to Ecosystem Service Valuation

There is a considerable demand for valuations of ecosystem services and natural capital from those who seek to argue that these receive insufficient attention from policy-makers and believe (perhaps correctly) that being able to put a 'price tag' on ecosystems will strengthen their argument. Indeed, several such 'total valuations' have been conducted, the most famous being that by Costanza *et al.*,[11] and the UK's Department for the Environment has commissioned economists to value the nation's ecosystem services.[80] Notwithstanding the potential propaganda value of these valuations, the desire to value ecosystems and their services misunderstands the nature and purpose of valuation. First, as has been pointed out,[9] it is the effects on human welfare that are valued, not the ecosystems themselves or their services, and these effects are produced through combining ecosystem services (or natural capital) and human or physical capital. Second, it is the incremental change in the ecosystem service or natural capital which is valued, not the ecosystem service itself, and only at the smallest of scales will the two be identical. As has been pointed out,[81] any attempt to estimate the "total value of the world's ecosystem services and natural capital" (as in ref. 10) would be a "serious underestimate of infinity", and a similar criticism could be levelled at total valuations of a nation's ecosystem services. Finally, economic valuation is predominantly concerned with the effects on human welfare of specific and plausible human actions, which may affect ecosystems, the services they provide, or the way these services are used. It is really the human action or intervention which is valued, not the ecosystem or ecosystem services which it affects. Thus, instead of valuing soils *per se,* what we can and should do is value the costs and benefits of actions which we might plausibly take which will improve (or degrade) our soil, in order to determine whether the action is desirable. Valuations of this kind are difficult for soil protection actions since, as noted above, soil change affects human wellbeing through many intermediate processes, many of which are imperfectly understood by science. As a consequence, they remain rare and good information on the net benefits of soil protection is severely lacking in both developed and developing countries.

Acknowledgements

Funding for David Robinson and Brian Reynolds was provided in part by the European Commission FP 7 Collaborative Project 'Soil Transformations in European Catchments' (SoilTrEC) (Grant Agreement No. 244118).

References

1. V. M. D'Costa, K. M. McGrann, D. W. Hughes and G. D. Wright, *Science*, 2006, **311**, 374–377.
2. C. Parkinson, Soil bacterium helps kill cancers, 2011, http://www.bbc.co.uk/news/health-14761417, accessed 12/04/2012.
3. MEA, Millennium Ecosystem Assessment, *Ecosystems and Human Well-being: Synthesis*, Island Press, Washington DC, 2005.
4. W. Westman, *Science*, 1977, **197**, 960–964.
5. G. Daily, *Natures Services: Societal Dependence on Natural Ecosystems*, Island Press, Washington DC, 1997.
6. R. Costanza, J. Cumberland, H. Daly, R. Goodland and R. Norgaard, *An Introduction to Ecological Economics*, St. Lucie Press, Boca Raton, FL, 1997.
7. K. J. Wallace, *Biol. Conserv.*, 2007, **139**, 235–246.
8. J. Boyd and S. Banzhaf, *Ecol. Econ.*, 2007, **63**, 616–626.
9. B. Fisher and R. K. Turner, *Biol. Conserv.*, 2008, **141**, 1167–1169.
10. B. Fisher, R. K. Turner and P. Morling, *Ecol. Econ.*, 2009, **68**, 643–653.
11. R. Costanza, R. d'Arge, R. deGroot, S. Farber, M. Grasso, B. Hannon, K. Limburg, S. Naeem, R. V. Oneill, J. Paruelo, R. G. Raskin, P. Sutton and M. vandenBelt, *Nature*, 1997, **387**, 253–260.
12. NEA, *The UK National Ecosystem Assessment Technical Report*, UNEP-WCMC, Cambridge, 2011.
13. E. Schumacher, *Small is Beautiful: A Study of Economics As If People Mattered*, Blond & Briggs, London, 1973.
14. Anon, The House of Commons, 1837, vol. 24, p. 109.
15. R. Costanza and H. Daly, *Conserv. Biol.*, 1992, **6**, 37–46.
16. G. Daily, P. Matson and P. Vitousek, in *Nature's Services: Societal Dependence on Natural Ecosystems*, ed. G. Daily, Island Press, Washington DC, 1997.
17. S. S. Andrews, D. L. Karlen and C. A. Cambardella, *Soil Sci. Soc. Am. J.*, 2004, **68**, 1945–1962.
18. D. Wall, *Sustaining Biodiversity and Ecosystem Services in Soils and Sediments*, Island Press, Washington DC, 2004.
19. S. M. Swinton, F. Lupi, G. P. Robertson and S. K. Hamilton, *Ecol. Econ.* 2007, **64**, 245–252.
20. C. Palm, P. Sanchez, S. Ahamed and A. Awiti, in *Annual Review of Environment and Resources*, 2007, vol. 32, pp. 99–129.
21. D. A. Robinson, I. Lebron and H. Vereecken, *Soil Sci. Soc. Am. J.*, 2009, **73**, 1904–1911.
22. E. Dominati, M. Patterson and A. Mackay, *Ecol. Econ.*, 2010, **69**, 1858–1868.
23. K. Bristow, S. Marchant, M. Deurer and B. Clothier, Soil solutions for a changing world, *Proceedings of the 19th World Congress of Soil Science, Brisbane, QLD, Australia*, 2010.
24. D. Robinson, N. Hockley, E. Dominati, I. Lebron, K. Scow, B. Reynolds, B. Emmett, A. Keith, L. de Jonge, P. Schjønning, P. Moldrup, S. Jones and M. Tuller, *Vadose Zone J.*, 2012, **11**, 1, doi:10.2136/vzj2011.0051.

25. B. Emmett, B. Reynolds, P. Chamberlain, E. Rowe, D. Spurgeon, S. Brittain, Z. Frogbrook, S. Hughes, A. Lawlor, J. Poskitt, E. Potter, D. Robinson, A. Scott, C. Wood and C. Woods, *Countryside Survey: Soils Report from 2007*, Centre for Ecology and Hydrology, Wallingford, Oxon UK, 2010.
26. H. Jenny, *Factors of Soil Formation: A System of Quantitative Pedology*, Dover Press, New York, 1994.
27. D. D. Richter, A. R. Bacon, M. L. Mobley, C. J. Richardson, S. S. Andrews, L. West, S. Wills, S. Billings, C. A. Cambardella, N. Cavallaro, J. E. DeMeester, A. J. Franzluebbers, A. S. Grandy, S. Grunwald, J. Gruver, A. S. Hartshorn, H. Janzen, M. G. Kramer, J. K. Ladha, K. Lajtha, G. C. Liles, D. Markewitz, P. J. Megonigal, A. R. Mermut, C. Rasmussen, D. A. Robinson, P. Smith, C. A. Stiles, R. L. Tate, A. Thompson, A. J. Tugel, H. van Es, D. Yaalon and T. M. Zobeck, *Soil Sci. Soc. Am. J.*, 2011, **75**, 2079–2084.
28. G. Certini and R. Scalenghe, *Soils. Basic Concepts and Future Challenges*, Cambridge University Press, Cambridge, UK, 2006.
29. N. Ramankutty, A. T. Evan, C. Monfreda and J. A. Foley, *Global Biogeochem. Cycles*, 2008, **22**, GB1003, doi:10.1029/2007GB002952.
30. S. Siebert, P. Doll, J. Hoogeveen, J. M. Faures, K. Frenken and S. Feick, *Hydrol. Earth Syst. Sci.*, 2005, **9**, 535–547.
31. J. A. Foley, N. Ramankutty, K. A. Brauman, E. S. Cassidy, J. S. Gerber, M. Johnston, N. D. Mueller, C. O'Connell, D. K. Ray, P. C. West, C. Balzer, E. M. Bennett, S. R. Carpenter, J. Hill, C. Monfreda, S. Polasky, J. Rockstrom, J. Sheehan, S. Siebert, D. Tilman and D. P. M. Zaks, *Nature*, 2011, **478**, 337–342.
32. G. P. Robertson, E. A. Paul and R. R. Harwood, *Science*, 2000, **289**, 1922–1925.
33. E. J. Kladivko, *Soil Tillage Res.*, 2001, **61**, 61–76.
34. L. B. Guo and R. M. Gifford, *Global Change Biol.*, 2002, **8**, 345–360.
35. M. D. Flowers and R. Lal, *Soil Tillage Res.*, 1998, **48**, 21–35.
36. R. Lal, *Environ. Int.*, 2003, **29**, 437–450.
37. D. A. Robinson, C. S. Campbell, J. W. Hopmans, B. K. Hornbuckle, S. B. Jones, R. Knight, F. Ogden, J. Selker and O. Wendroth, *Vadose Zone J.*, 2008, **7**, 358–389.
38. G. Schoups, J. W. Hopmans, C. A. Young, J. A. Vrugt, W. W. Wallender, K. K. Tanji and S. Panday, *Proc. Natl. Acad. Sci. U. S. A.*, 2005, **102**, 15352–15356.
39. J. A. Foley, R. DeFries, G. P. Asner, C. Barford, G. Bonan, S. R. Carpenter, F. S. Chapin, M. T. Coe, G. C. Daily, H. K. Gibbs, J. H. Helkowski, T. Holloway, E. A. Howard, C. J. Kucharik, C. Monfreda, J. A. Patz, I. C. Prentice, N. Ramankutty and P. K. Snyder, *Science*, 2005, **309**, 570–574.
40. J. N. Galloway, J. D. Aber, J. W. Erisman, S. P. Seitzinger, R. W. Howarth, E. B. Cowling and B. J. Cosby, *Bioscience*, 2003, **53**, 341–356.

41. J. Rockstrom, W. Steffen, K. Noone, A. Persson, F. S. Chapin, E. F. Lambin, T. M. Lenton, M. Scheffer, C. Folke, H. J. Schellnhuber, B. Nykvist, C. A. de Wit, T. Hughes, S. van der Leeuw, H. Rodhe, S. Sorlin, P. K. Snyder, R. Costanza, U. Svedin, M. Falkenmark, L. Karlberg, R. W. Corell, V. J. Fabry, J. Hansen, B. Walker, D. Liverman, K. Richardson, P. Crutzen and J. A. Foley, *Nature*, 2009, **461**, 472–475.
42. R. J. Haynes and R. Naidu, *Nutr. Cycling Agroecosyst.*, 1998, **51**, 123–137.
43. S. Duahyentumi and D. B. Johnson, *Soil Biol. Biochem.*, 1986, **18**, 629–635.
44. S. Angel, S. C. Sheppard and D. L. Civco, *The Dynamics of Global Urban Expansion*, Transport and Urban Development Department, The World Bank, Washington DC, 2005.
45. R. Scalenghe and F. A. Marsan, *Landscape Urban Plann.*, 2009, **90**, 1–10.
46. D. Fowler, K. Pilegaard, M. A. Sutton, P. Ambus, M. Raivonen, J. Duyzer, D. Simpson, H. Fagerli, S. Fuzzi, J. K. Schjoerring, C. Granier, A. Neftel, I. S. A. Isaksen, P. Laj, M. Maione, P. S. Monks, J. Burkhardt, U. Daemmgen, J. Neirynck, E. Personne, R. Wichink-Kruit, K. Butterbach-Bahl, C. Flechard, J. P. Tuovinen, M. Coyle, G. Gerosa, B. Loubet, N. Altimir, L. Gruenhage, C. Ammann, S. Cieslik, E. Paoletti, T. N. Mikkelsen, H. Ro-Poulsen, P. Cellier, J. N. Cape, L. Horvath, F. Loreto, U. Niinemets, P. I. Palmer, J. Rinne, P. Misztal, E. Nemitz, D. Nilsson, S. Pryor, M. W. Gallagher, T. Vesala, U. Skiba, N. Brueggemann, S. Zechmeister-Boltenstern, J. Williams, C. O'Dowd, M. C. Facchini, G. de Leeuw, A. Flossman, N. Chaumerliac and J. W. Erisman, *Atmos. Environ.*, 2009, **43**, 5193–5267.
47. F. Dentener, J. Drevet, J. F. Lamarque, I. Bey, B. Eickhout, A. M. Fiore, D. Hauglustaine, L. W. Horowitz, M. Krol, U. C. Kulshrestha, M. Lawrence, C. Galy-Lacaux, S. Rast, D. Shindell, D. Stevenson, T. Van Noije, C. Atherton, N. Bell, D. Bergman, T. Butler, J. Cofala, B. Collins, R. Doherty, K. Ellingsen, J. Galloway, M. Gauss, V. Montanaro, J. F. Muller, G. Pitari, J. Rodriguez, M. Sanderson, F. Solmon, S. Strahan, M. Schultz, K. Sudo, S. Szopa and O. Wild, *Global Biogeochem. Cycles*, 2006, **20**, GB4003, doi:10.1029/2005GB002672.
48. L. A. Derry and O. A. Chadwick, *Elements*, 2007, **3**, 333–338.
49. I. Tegen, M. Werner, S. P. Harrison and K. E. Kohfeld, *Geophys. Res. Lett.*, 2004, **31**, L05105, doi:10.1029/2003GL019216.
50. N. M. Mahowald, A. R. Baker, G. Bergametti, N. Brooks, R. A. Duce, T. D. Jickells, N. Kubilay, J. M. Prospero and I. Tegen, *Global Biogeochem. Cycles*, 2005, **19**, GB4025, doi:10.1029/2004GB002402.
51. G. R. Walther, E. Post, P. Convey, A. Menzel, C. Parmesan, T. J. C. Beebee, J. M. Fromentin, O. Hoegh-Guldberg and F. Bairlein, *Nature*, 2002, **416**, 389–395.
52. J. H. Brown, T. J. Valone and C. G. Curtin, Proc. Natl. Acad. Sci. U. S. A., 1997, **94**, 9729–9733.
53. S. I. Seneviratne, D. Luthi, M. Litschi and C. Schar, *Nature*, 2006, **443**, 205–209.

54. J. Walker, F. Bullen and B. G. Williams, *J. Appl. Ecol.*, 1993, **30**, 265–273.
55. S. T. Berthrong, E. G. Jobbagy and R. B. Jackson, *Ecol. Appl.*, 2009, **19**, 2228–2241.
56. I. Lebron, D. A. Robinson, M. Oatham and M. N. Wuddivira, *J. Hydrol.*, 2012, **414–415**, 194–200.
57. S. H. Doerr, R. A. Shakesby and R. P. D. Walsh, *Earth-Sci. Rev.*, 2000, **51**, 33–65.
58. D. A. Wardle, R. D. Bardgett, J. N. Klironomos, H. Setala, W. H. van der Putten and D. H. Wall, *Science*, 2004, **304**, 1629–1633.
59. P. H. Bellamy, P. J. Loveland, R. I. Bradley, R. M. Lark and G. J. D. Kirk, *Nature*, 2005, **437**, 245–248.
60. L. Martino and M. Fritz, *New Insight into Land Cover and Land Use in Europe; Land Use/Cover Area Frame Statistical Survey: Methodology and Tools*, http://epp.eurostat.ec.europa.eu/cache/ITY_OFFPUB/KS-SF-08-033/EN/KS-SF-08-033-EN.PDF, 2008, accessed 12/04/2012.
61. Anon, *Biosoil*, http://forest.jrc.ec.europa.eu/contracts/biosoil. 2011, accessed 12/04/2012
62. P. D. Carey, S. Wallis, P. M. Chamberlain, A. Cooper, B. A. Emmett, L. C. Maskell, T. McCann, J. Murphy, L. R. Norton, B. Reynolds, W. A. Scott, I. C. Simpson, S. M. Smart and J. M. Ullyett, *Countryside Survey: UK Results from 2007*, 2008.
63. L. G. Firbank, C. J. Barr, R. G. H. Bunce, M. T. Furse, R. Haines-Young, M. Hornung, D. C. Howard, J. Sheail, A. Sier and S. M. Smart, *J. Environ Manage.*, 2003, **67**, 207–218.
64. P. Smith, S. J. Chapman, W. A. Scott, H. I. J. Black, M. Wattenbach, R. Milne, C. D. Campbell, A. Lilly, N. Ostle, P. E. Levy, D. G. Lumsdon, P. Millard, W. Towers, S. Zaehle and J. U. Smith, *GCB Bioenergy*, 2007, **13**, 2605–2609.
65. R. I. Griffiths, B. C. Thomson, P. James, T. Bell, M. Bailey and A. S. Whiteley, *Environ. Microbiol.*, 2011, **13**, 1642–1654.
66. RoTAP, *Review of Transboundary Air Pollution, Acidification, Eutrophication, Ground Level Ozone and Heavy Metals in the UK*, Defra, UK, 2010.
67. G. J. D. Kirk, P. H. Bellamy and R. M. Lark, *GCB Bioenergy*, 2010, **16**, 3111–3119.
68. I. Lebron, A. M. Keith, S. Hughes, B. Reynolds, D. A. Robinson, D. M. Cooper and B. A. Emmett, *Evaluation of the Occurrence of Soils with pH Higher than 8.4 Observed within the Countryside Survey*, Project SP1304, NERC/Centre for Ecology and Hydrology, 2011.
69. D. S. LeBauer and K. K. Treseder, *Ecology*, 2008, **89**, 371–379.
70. P. W. Hill, J. Farrar, P. Roberts, M. Farrell, H. Grant, K. K. Newsham, D. W. Hopkins, R. D. Bardgett and D. L. Jones, *Nature Climate Change*, 2011, **1**, doi: 10.1038/NCLIMATE1060.
71. E. C. Rowe, Z. L. Frogbrook, S. Hughes, D. A. Robinson and B. A. Emmett, *Sci. Total Environ.*, 2011, doi:10.1016/j.scitotenv.2011.12.027.

72. J. Hall, M. Ashmore, J. Fawehinimi, C. Jordan, S. Lofts, L. Shotbolt, D. Spurgeon, C. Svendsen and E. Tipping, *Environ. Toxicol. Chem.*, 2006, **23**, 833–890.
73. S. Lofts, D. Spurgeon and C. Svendsen, *Environ. Sci. Technol.*, 2005, **39**, 8533–8540.
74. L. R. Lado, T. Hengl and H. I. Reuter, *Geoderma*, 2008, **148**, 189–199.
75. S. P. McGrath and P. J. Loveland, *The Soil Geochemical Atlas of England and Wales.*, Blackie Academic and Professional, Glasgow, 1992.
76. Anon, ed. P. B. Department of Environment, London E15 2HF London: HMSO, 1994.
77. M. Rutgers, A. J. Schouten, J. Bloem, N. van Eekeren, R. G. M. de Goede, G. Akkerhuis, A. van der Wal, C. Mulder, L. Brussaard and A. M. Breure, *Eur. J. Soil Sci.*, 2009, **60**, 820–832.
78. J. Bengtsson, *J. Animal Ecol.*, 1994, **63**, 653–665.
79. C. Bullock, C. Kretsch and E. Candon, *The Economic and Social Aspects of Biodiversity –Benefits and Costs of Biodiversity in Ireland.*, The Stationary Office, Government of Ireland, 2008.
80. Anon, *An Introductory Guide to Valuing Ecosystem Services*, PB12852, Department for Environment, Food and Rural Affairs, London, 2007.
81. M. Toman, *Ecol. Econ.*, 1998, **25**, 57–60.

The Evaluation and Reporting of Soils in Sustainable Agriculture and Food Systems

ALFRED E. HARTEMINK*[a], AMANDA L. RASTER[b] AND MOLLY M. JAHN[c]

[a] Department of Soil Science, FD Hole Soils Lab, University of Wisconsin, 1525 Observatory Drive, Madison, WI 53706, USA; [b] Department of Horticulture, University of Wisconsin, 1575 Linden Drive, Madison, WI 53706, USA; [c] Laboratory of Genetics and Department of Agronomy, University of Wisconsin, 425 Henry Mall, Madison, WI 53706, USA
*E-mail: hartemink@wisc.edu or alraster@wisc.edu or mjahn@wisc.edu

ABSTRACT

Sustainable agriculture has been on the global agenda for several decades now. Because soil is of critical importance in meeting the goals and objectives of sustainable agriculture, there is a need to assess and evaluate the soil under different agricultural and food systems. Here we present an overview of measurement tools and reporting standards that are currently being used in the USA and globally. We describe three measurement tools used primarily in the USA, followed by a review of eight USA standards and twelve standards that have international applications. These standards and metrics encompass a range of sustainability criteria such as soil and water conservation and quality, crop nutrient management, energy conservation and greenhouse gas

Issues in Environmental Science and Technology No. 35
Soils and Food Security
Edited by R.E. Hester and R.M. Harrison

Published by the Royal Society of Chemistry, www.rsc.org

emissions, biological diversity, integrated pest management and reduced pesticide risk, safe and fair working conditions and economic viability for agricultural producers. They range from private to multi-stakeholder in terms of inclusivity during development, whole farm to crop-specific in scope, practice-based to outcomes-based in implementation and self-assessment to certification to consumer-facing eco-label in terms of verification. From this review, we conclude that this first set of evaluation, monitoring and reporting protocols and standards (Tier 1) allow for a second Tier of sustainability that focuses on the use of metrics and other measures of the impacts of agriculture toward reducing the environmental footprint of agriculture, identifying synergies through innovation and improving critical outcomes in human dimensions. We propose the concept of Tier 3 sustainability whereby we commit to actions in agriculture and food systems that, in aggregate, will move us into "safe operating space" for human beings and our planet.

1 Introduction

Soil science has always had strong ties with agriculture, and soil science knowledge has made large contributions to the increase in agricultural production A better understanding of soils has been essential for research questions on climate change, environmental regulation, crop productivity and nutrient management in agricultural and ecosystem services. We begin the 21st century with an energised recognition of the role that soil and soil science will play in the renovation of agriculture to better meet human needs within planetary safe operating space.[1] We also begin the 21st century suffering the effects of reduced funding for soil science, and agricultural research more generally, that started in many countries in the mid-1980s; this followed widespread governmental budget cuts and a reduced interest in agriculture.

Today there is widespread concern about the land base necessary to meet current and future energy (biofuels), food (hunger alleviation, increasing population) and feed (increasing animal production) needs. High oil prices contribute to greater demand for biofuels. In some parts of the world, the cultivation of biofuel crops is competing with food crops and driving up commodity prices.[2] Although the environmental – and particularly soil – impact of the shift towards growing crops for energy and increased food production is still under assessment, it is widely realised that global soil information is not accurate nor digitally available and is not up-to-date.[3]

After the Bruntland report of 1987 on *Our Common Future* and the Earth Summit in Rio de Janeiro in 1992, there has been a global debate on the issue of sustainability, particularly in relation to soils, land and agriculture. The end of the 20th century will be remembered as an era in which the term 'sustainability' was overwhelmingly present in the agriculture and soil science literature. However, evidence that this discussion has produced positive impacts on either agriculture or the stability of the Earth system is scant at best. Indeed, many negative trends, such as the loss of agricultural lands to urbanisation and degradation of soils in agricultural regions continues[4–7] and have dramatically accelerated agricultural policymaking since the sustainability debate emerged.

Overall, while sustainability is an inherently vague concept whose scientific definition and measurement lack broad agreement,[8] virtually any accepted definition includes the recognition that sustainable practices protect future generations' access to essential ecosystems goods and services. Although 'sustainability' is widely used in economic and development contexts, here we will emphasise 'sustainability' in its biophysical and ecological context, with an emphasis on the sustainability of soil resources for agricultural production. In past decades, a plethora of definitions of sustainable land management has been produced.[9] Some have been very lengthy, but, in essence, a biophysical definition of sustainability refers to the combination of production and conservation of the natural resources on which the production depends.[10] Furthermore, sustainable agricultural production should not release any products that make the environment less desirable for human occupation and which cannot readily be removed.[9] In a wider sense, agricultural sustainability will be realised in a system that provides nutritional security and a healthy diet for all under conditions that will allow us to continue to meet these demands in the future.

The soil is the most important component in sustainable land management, which has been indicated by pedologists, soil fertility experts and soil biologists.[11–13] Sustainable land management is deemed necessary for both the developed world, where high external input agriculture dominates, and for the less developed countries, where the agricultural production is locally dominated by low yields, little or no nutrient inputs, and inadequate soil conservation practices. Sustainability problems occur through under-use of resources (soil degradation) or over-use of resources (environmental effects of nutrients and pesticides). Crop yields are key indicators for assessing sustainable land management, but other indicators, as well as the determination of threshold values at which crop productivity is affected, are also important.

Every approach to the challenge of describing and defining sustainable agriculture has included soil characteristics and management in one way or another. Because of the intersection of soils and sustainable agriculture, we have undertaken a review of the ways in which soils are currently reflected in sustainability reporting as applied to agriculture and food systems. In three sections, we review the ways in which a number of major agricultural sustainability initiatives that include both practice-based approaches (where specific practices are required to meet the standard) and metrics-based approaches (where producers implement their choice of best management practices in order to meet specific outcomes) address soil management. Firstly, three measurement tools are described, followed by a review of eight US standards and twelve standards that are being developed, or have been developed, globally. Both practice-based and metrics-based approaches that address soil management are reviewed, followed by a general discussion of how these standards and protocols can help us to further enhance the sustainable agriculture paradigm. This chapter is offered at a crucial moment when there is increasing global attention for both soils and soil research,[14] part of the renewed global interest in agriculture as we approach a population of nine billion by 2050.

2 Measurement Tools

2.1 Field to Market Initiative for Sustainable Agriculture

Led by the Keystone Center, a nonprofit US organisation that uses a consensus-based approach to address issues in energy, environment, health and education, Field to Market is a multi-stakeholder initiative that aims to create targeted sustainability outcomes for USA agriculture. Field to Market has developed a series of indicators[15] for estimating land use, irrigation water use, energy use, greenhouse gas emissions, soil loss and soil carbon and continues to work on the development of metrics to estimate additional on-farm environmental, economic, social and health outcomes related to the production of corn, soy, wheat, cotton and rice.

Field to Market's Fieldprint Calculator is an online tool designed to help producers assess how their crop production practices impact the sustainability of their farming operations. The calculator generates a "fieldprint" value that producers can compare to county, state and national averages. In terms of soil parameters, the tool focuses on soil loss and soil carbon. The Soil Conservation Resource metric accounts for soil losses due to wind and water erosion, which is measured in units of soil loss in weight per year per unit of production. The metric is based on the Revised Universal Soil Loss Equation 2 (RUSLE2) model and the Wind Erosion Prediction System 1.0 (WEPS 1.0). Producers are asked to enter information on field characteristics (such as soil texture, soil erodibility, slope and slope length), tillage and crop management practices, and whether or not structures like drainage and wind barriers or terraces are in place. Based on these inputs, the models return values for soil losses due to water and wind, as well as tolerable (T) soil loss due to water. The calculator also quantifies soil carbon using a relative measure called the Soil Conditioning Index (SCI). The SCI is an output of the RUSLE2 model and is determined by the amount of organic matter added to the soil, field operations and soil erosion rates. The soil carbon metric takes on a value of -1 to $+1$. As the value moves in a positive or negative direction away from zero, the magnitude of the number represents the confidence in soil carbon either being added or removed from the soil. The SCI does not measure the rate of change in soil carbon nor is it specific to a given crop.

2.2 Stewardship Index for Specialty Crops

The Stewardship Index for Specialty Crops (SISC) is a multi-stakeholder initiative working to develop a system for measuring sustainable performance throughout the specialty crop supply chain. The effort aims to establish a common suite of outcomes-based metrics[16] to enable operators at any point along the supply chain to benchmark, compare and communicate their performance in meeting sustainability goals. Currently, there are six metrics in the pilot-testing phase, which fall under four categories: Energy, Soil, Nutrients and Water.

The SISC Soil metric focuses on soil organic matter (SOM), the organic fraction of the soil, excluding non-decomposed plant and animal residues. The amount of total organic carbon (TOC) present in the soil – which is used as an indicator of SOM – is divided by the soil's potential to store organic carbon, which is determined by the US Department of Agriculture's (USDA) Soil Management Assessment Framework (SMAF).[17] Increasing amounts of SOM, reflected as an increase in the value of the metric, signifies the soil's ability to improve nutrient delivery to plants, retain water, drain excess water and resist disease and erosion. To normalise against variability due to climate, soil type and soil texture, SOM is compared with a site-specific estimate of the soil's potential to hold SOM.

2.3 Cool Farm Tool

Developed by the University of Aberdeen, the Cool Farm Tool[18] is a farm-level greenhouse gas calculator that estimates net greenhouse gas emissions from agriculture. Two of the calculator's six input sections, which include: Crop Management, Sequestration, Livestock, Field Energy Use, Processing and Transport, cover soil parameters.

Under the Crop Management section, producers are asked to report on the soil texture, soil moisture, soil organic matter, drainage capacity and pH of the field under analysis. Producers must also identify which fertilisers they use, method and rate of fertiliser application, whether or not compost and/or emissions inhibitors are used and how much crop residue is managed post-harvest. The intent of the questions is to estimate emissions (primarily nitrous oxide; N_2O) from the soil and from crop residue.

In the Sequestration section, producers report on land use and farm management changes, such as conversion of grassland to arable production, forest clearing, changes in tillage methods, cover cropping, compost and residue incorporation. Where applicable, producers may also enter information on annual biomass accumulated or removed from productive tree species in or near the field. The cumulative total of the Crop Management and Sequestration sections indicates the amount of soil C accumulated or lost from the production system and provides producers with an assessment of how management and land use changes affect the operation's overall greenhouse

Table 1 Summary of measurement tools relating soils and sustainable agriculture.

Measurement tool	*Crop(s)*	*Soil properties*	*Assessment method(s)*
Field to Market Initiative for Sustainable Agriculture	Corn, cotton, rice, soy, wheat	Soil erosion potential, soil carbon	Revised Universal Soil Loss Equation 2 (RUSLE2); Wind Erosion Prediction System 1.0 (WEPS 1.0)
Stewardship Index for Specialty Crops	Vegetables, fruits, nuts, horticulture	Soil organic matter, total organic carbon	Total organic carbon present in soil / carbon storage potential (determined by USDA's Soil Management Assessment Framework)
Cool Farm Tool	Grain, legume, fruit and vegetable crops; livestock	Soil moisture, soil type, soil organic matter, soil drainage, pH, soil carbon	Soil and foliage tests

gas emissions. Table 1 summarises the three measurement tools discussed above.

3 Sustainability Schemes in the USA

The following section summarises select sustainability schemes currently used by agricultural producers in the USA to document farming practices for purposes of benchmarking, verification, certification and continual improvement over time. Although the following schemes differ in scope, types of production systems assessed and mode of implementation, there is extensive overlap in the specific soil properties addressed and the methods used to monitor and/or quantify them.

3.1 California Almond Board Sustainability Program

The California Almond Sustainability Program,[19,20] developed by the California Almond Board, is a self-assessment tool used by almond growers to benchmark their farming practices against a range of best management practices and other measures of farm efficiency. The assessment currently consists of three modules: Irrigation Management, Nutrient Management and Energy Efficiency.

The Irrigation Management[19] module guides growers through the process of determining irrigation volume and timing based on key soil parameters. The tool provides information on available water for different soil types (coarse, sandy, medium, fine), typical efficiencies of various types of irrigation systems (where efficiency is equal to water stored/water applied) and how to measure soil and water salinity. These values are then used to determine the soil's leaching fraction (*i.e.* the amount of water needed to flush salts beyond the plant root zone). Additionally, growers are asked to indicate the water penetration capabilities of the soil under assessment, soil pH and levels of soil organic matter (SOM), and to identify any practices implemented to improve overall soil quality. The grower then uses this information to assist in irrigation scheduling and to determine appropriate volumes of irrigation water to apply to the crop.

The Nutrient Management Module[20] helps growers to assess soil organic matter, soil pH, variations in soil characteristics that might affect plant nutrient availability and uptake, and the contribution of various nutrient sources to the overall nutrient budget in order to maximise nutrient use efficiency.

3.2 Council on Sustainable Biomass Production Draft Provisional Standard for Sustainable Production of Agricultural Biomass

The Council on Sustainable Biomass Production (CSBP) is a multi-stakeholder organisation working on the development of a comprehensive voluntary standard[21] for sustainable production of biomass and its conversion to bioenergy. The standard is intended to serve as a foundation for an independent, third-party certification program aimed at setting the emerging bioenergy industry on a course of continuously improving best practices related to sustainability. Producers conduct a preliminary self-assessment of their compliance to the standard, using a third-party auditor for quality assurance. The baseline evaluation provides the grower with a comprehensive report on areas of compliance, as well as areas where improvement is needed in order to meet certification requirements.

The CSBP standard consists of nine principles that express key elements of sustainable biomass production: Integrated Resource Management Planning, Soil, Biological Diversity, Water, Climate Change, Socio-Economic Well-Being, Legality, Transparency and Continuous Improvement. The Soil principle is intended to demonstrate the importance of soil stability, soil fertility and organic matter to sustainable production. At the minimum, producers are required to take measures to minimise soil erosion, maintain carbon and nutrients at levels appropriate for biomass crop production and preserve the overall physical, chemical and biological properties of the soil. To meet the first level of certification, producers must conduct a soil assessment and base crop management decisions on soil characteristics and capabilities. Additionally, producers are required to use planning protocols supported by the Natural Resource Conservation Service (NRCS) for soil nutrient and conservation planning, to score less than or equal to tolerable (T) soil loss due to water on RUSLE2, to retain biomass materials on the landscape for erosion control and soil fertility, and to minimise soil compaction. At the second level of certification, producers must establish a comprehensive management plan and implement practices to improve soil function and productivity.

3.3 Demeter Biodynamic Farm Standard

The Demeter Association's Biodynamic Farm Standard[22] encourages producers to utilise the principles of living organisms in the management of their farming operations. The ultimate goal is to create a farming system that meets its needs from the living dynamics of the farm itself, and depends only minimally on off-farm materials. Soil fertility management plays a key role within the standard toward meeting this objective. Producers are required to incorporate practices that build soil humus and promote soil biological activity through the recycling of raw organic materials generated on the farm. Supported practices include using legumes in crop rotation, recycling livestock manures, incorporating green manure into the production system, and using biodynamic compost preparations. Several types of fertiliser materials, such as raw manure, biosolids and all forms of synthetic fertilisers, are restricted so as to minimise adverse impacts on soil biology. The standard requires the use and timely application of nine types of biodynamic preparations aimed toward revitalising the soil and stimulating plant root growth, enhancing development of microorganisms and humus formation and promoting photosynthetic activity.

3.4 Food Alliance Whole Farm/Ranch Inspection Tool

The Food Alliance Whole Farm/Ranch Inspection Tool[23] is a self-assessment tool that contains both fixed and variable evaluation criteria intended to guide crop and animal producers toward Food Alliance certification. Producers self-evaluate their production practices under four areas of sustainability – Reducing Pesticide Usage, Soil and Water Conservation, Safe and Fair Working Conditions and Wildlife Habitat – and tabulate the composite score to determine whether they quality for certification. Alternatively, third-party inspectors may walk producers through the assessment. Evaluation criteria are categorised according to four levels, with producers earning more points for successive levels of achievement.

The Soil Conservation component of the standard is paired with Water Conservation and focuses on such elements as implementation and management of buffer strips around waterways, soil erosion prevention, tillage practices, soil compaction prevention, nutrient and soil organic matter management and continuing education for soil and water conservation. At a minimum, producers are required to meet all applicable federal, state and local legal requirements related to erosion control. Additional points are earned for regularly monitoring soil erosion potential, utilising a tillage regime that conserves soil, maintaining soil pH to ensure proper nutrient availability and uptake, implementing a nutrient management plan and actively managing soil organic matter. Producers earn maximum points for retaining crop residue, relying entirely on no-till cultivation and increasing soil organic matter content in arable fields.

3.5 Wisconsin Vegetable Sustainability Standards/Healthy Farms Whole Farm Self-Assessment

The Wisconsin Whole Farm Vegetable Sustainability Standard[24] was developed by the University of Wisconsin, in collaboration with the World Wildlife Fund and the Wisconsin Potato and Vegetable Growers Association. Initially launched as a single-crop, single-state program to pursue more environmentally benign ways to grow fresh market potatoes in Wisconsin, the effort has since expanded to include a whole farm certification component, as well as crop-specific standards for multiple vegetable and rotational crops.

Under the Whole Farm assessment, growers are asked to identify their compliance to criteria under ten areas of sustainability: Ecosystem Restoration, Farm Production Management, Soil and Water Quality, Scouting, Information Sources, Pest Management Decisions, Pest Management, Resistance Management, Chain of Custody, and Farm Operations and Sustainability. Soil fertility practices are addressed under the Soil and Water Quality section, which asks growers to report on their nutrient management strategies relative to soil nutrient test results, management of soil pH and efforts to increase soil moisture holding capacity. Growers also earn points for preventing soil compaction and soil erosion and for implementing a soil conservation plan.

3.6 Lodi Rules for Sustainable Winegrowing

The Lodi Rules for Sustainable Winegrowing,[25] developed by the Lodi-Woodbridge Winegrape Commission, is a third-party certification program that guides California winegrape producers toward practices that improve soil, air and water quality and the overall sustainability of their farming operations. The assessment encourages a systems perspective that involves consideration of multiple factors when making field and farm-level decisions. Key areas of evaluation include: Ecosystem Management, Soil Management, Water Management, Vineyard Establishment, Pest Management, and Employee Education, Training and Teambuilding.

The Soil Management component of the Lodi Rules requires as a baseline the development of a comprehensive nutrition management plan that addresses soil analysis, field characteristics, sources and forms of nutrients and other relevant factors. Producers are asked to report on additional soil parameters, such as whether they have mapped the soil series of the vineyard under assessment, how frequently soils are sampled for micronutrient and macronutrient analysis, whether the application of soil amendments is modified according to irrigation water testing, whether the soil provides adequate nitrogen for winegrape production without application of supplemental N, and the implementation of any practices aimed to enhance soil water penetration, build soil organic matter and minimise soil erosion.

3.7 USDA National Organic Program

The National Organic Program (NOP)[26] is a regulatory program housed within the USDA Agricultural Marketing Service (USDA-AMS). The intent of the NOP is to assure consumers that products bearing the USDA Organic seal have been produced through approved methods that integrate cultural, biological and mechanical practices designed to promote resource cycling and ecological balance and conserve biodiversity. The USDA accredits certifying agents who are responsible for ensuring that certified farm and processing facilities produce products that meet or exceed all organic standards. Certified farms and processors are allowed to represent their products as organic.

Two sections of the organic standard address soil parameters: the Soil Fertility and Crop Nutrient Management Practice Standard and the Crop Rotation Practice Standard. The Soil Fertility and Crop Nutrient Management Practice Standard[27] requires producers to (1) implement tillage and cultivation practices that improve the physical, chemical and biological condition of soil and minimise soil erosion; (2) manage crop nutrients and soil fertility through rotations, cover crops and the application of approved plant and animal materials; and (3) maintain or improve soil organic matter content in a manner that does not contribute to contamination of crops, soil or water. Additionally, producers are prohibited from using biosolids and any fertilisers or composted materials that contain synthetic substances not included on the allowed materials list. Under the Crop Rotation Practice Standard,[28] producers are required to implement a crop rotation regime that maintains or improves soil organic matter content, manages deficient or excess plant nutrients and provides erosion control.

3.8 Central Coast Vineyard Sustainability in Practice Certification Program

The Sustainability in Practice Certification Program (SIP)[29] was developed by the Central Coast Vineyard Team to promote environmentally safe and economically sustainable farming methods among California winegrape growers. The SIP certification standards contain both required practices and management enhancements across ten sustainability areas, including Conservation and Enhancement of Biodiversity, Vineyard Establishment and Management, Soil Conservation and Water Quality, Water Conservation, Energy Conservation and Efficiency, Air Quality, Social Equity, Pest Management, Continuing Education, and Product Assurance and Business Sustainability. In addition, growers must develop a farm plan in order to qualify for certification.

The objective of the Soil Conservation component of the standard is to maintain healthy soils for optimal vine growth, development and production. To that end, the Soil Conservation requirements are aimed at helping growers understand their soil characteristics to the degree possible, to conserve and improve beneficial soil attributes, and use best management practices to correct deficiencies in soil tilth, water and nutrient status. At a minimum, growers must determine the soil series of the vineyard under assessment, as well as its erosion hazard, permeability and run-off rates in order to assist in production-related decision-making. Vineyard soils must also be tested at least every three years for nutrient content and monitored for pH, electrical conductivity (EC) and toxicities. Growers earn additional points for analysing the soil profile's physical and chemical characteristics, taking corrective action for alkaline, saline or acidic soils, building soil organic matter, minimising soil compaction, improving water infiltration, and eliminating erosion and offsite movement of sediment. Table 2 summarises the eight USA standards discussed above.

4 Global Sustainability Schemes

The following section summarises a subset of global sustainability schemes that are representative of the principles and criteria being assessed within a range of agricultural production systems across the world. As with the USA sustainability schemes, in spite of the differences in scope, production systems assessed, and mode of implementation, there is extensive overlap in the specific soil properties addressed and the methods used to monitor and/or quantify them.

4.1 Basel Criteria for Responsible Soy Production

The Basel Criteria for Responsible Soy Production,[30] developed by ProForest for Coop Switzerland and WWF Switzerland, provides soy producers with a working definition of environmentally, socially and economically responsible production. The criteria are designed to be applicable to all soy production at all scales throughout the world and cover the following aspects of sustainability: Legal Compliance, Technical Management, Environmental Management, Social Management, Continuous Improvement and Traceability. Soy producers can use the criteria as both an internal management tool and as a mechanism for confirming to buyers that their soy products originate from a responsibly managed source. Producers who wish to use the criteria as a market communication mechanism are required to demonstrate compliance through a third-party assessment.

Soil parameters are addressed under Technical Management. The criteria ask producers to assess soil suitability for soy cultivation by way of soil maps or soil surveys, and to use this information in their production plans. Efforts to maintain long-term soil fertility through cultural practices, such as appropriate fertiliser application and crop rotations, are recommended, as are measures to

Table 2 Summary of measurement tools relating soils and sustainable agriculture.

Standards	*Crop(s)*	*Soil properties*	*Assessment method(s)*
California Almond Sustainability Program	Almonds	Soil available water, soil type, permeability, soil salinity, leaching fraction, pH, soil organic matter	Soil maps/surveys, remote sensing technology, soil and foliage test, producer observation
CSBP Draft Provisional Standard for Sustainable Production of Agricultural Biomass	Biomass crops	Soil erosion, soil carbon, soil nutrients, soil type, soil structure; overall physical, chemical and biological properties	RUSLE2, NRCS planning protocols, soil maps/surveys, soil and foliage tests, soil conditioning index, producer observation
Demeter Biodynamic Farm Standard	Crops and livestock	Soil nutrients, soil organic matter, soil biological activity	Soil nutrient budgets, producer observation
Food Alliance Whole Farm/ Ranch Inspection Tool	Crops and livestock	Soil erosion potential, soil structure, soil nutrients, soil organic matter, soil pH	Soil and foliage tests, producer observation
Wisconsin Vegetable Sustainability Standards/ Healthy Farms Whole Farm Self-Assessment	Potatoes and processing vegetables	Soil nutrients, soil pH, soil moisture holding capacity, soil structure, soil erosion potential	Soil and foliage tests, producer observation
Lodi Rules for Sustainable Winegrowing	Winegrapes	Soil nutrients, soil type, soil structure, soil permeability, soil organic matter, soil erosion potential	Soil and foliage tests, soil maps/ surveys, producer observation
USDA National Organic Program	Crops and livestock	Soil erosion potential; soil nutrients; soil organic matter; overall physical, chemical and biological properties	None specified
Central Coast Vineyard Sustainability in Practice Certification Program	Winegrapes	Soil structure, soil type, soil nutrients, soil moisture holding capacity, soil erosion potential, soil permeability, pH, electrical conductivity, soil organic matter	Soil maps/surveys, soil and foliage tests, producer observation

minimise soil erosion and damage to soil structure. For example, the criteria encourage producers to use mechanical cultivation only where proven to improve or maintain soil structure and to avoid soil compaction.

4.2 Better Cotton Initiative

Founded to promote measurable improvements in the primary environmental and social impacts of cotton production worldwide, the Better Cotton Initiative (BCI) is a multi-stakeholder initiative comprised of organisations from across the cotton supply chain, NGOs and other interested parties. The BCI framework[31] contains Minimum Production Criteria that cotton producers must meet as part of a self-assessment process and Progress Requirements, of which a minimum must be met in order to qualify as a Better Cotton producer.

Impact areas covered under BCI include: Crop Protection, Water, Soil, Habitat, Fiber Quality and Decent Work. To meet the Minimum Production Criteria under Soil, producers must have knowledge of (1) appropriate soil management practices for identifying, preserving and enhancing soil structure and organic matter levels; (2) appropriate nutrient monitoring procedures, nutrient formulations and application timing and techniques; and (3) soil erosion management practices. Progress Requirements are met if producers actually implement soil management practices that preserve soil structure, increase soil organic matter content, address soil structural problems and control/prevent erosion. Additional Progress Requirements include applying nutrients on the basis of identified crop and soil need following a nutrient budget and regularly monitoring the nutrient status of the crop and soil and long-term nutrition trends in the area of production under assessment.

4.3 European Integrated Farming Framework

The European Integrated Farming Framework[32] contains guidelines, practices and suggestions for sustainable development in European agriculture. Drafted by the European Initiative for Sustainable Development in Agriculture (EISA), the Framework is intended to serve as a comprehensive management tool for farmers, to harmonise agricultural production across Europe, and to help create a better public and political understanding of integrated farming.

The Framework is organised into chapters that cover Organisation, Management and Planning; Human and Social Capital; Energy Use and Efficiency; Water Use and Protection; Air Emissions; Soil Management; Crop Nutrition; Crop Protection; Animal Husbandry and Health; Landscape, Wildlife and Biodiversity; and Resource Management, Product Storage and Waste Disposal. Soil parameters, such as soil organic matter, soil physical structure and sufficient fertility, are addressed under Soil Management and Crop Nutrition. Guidelines are organised into three tiers – Must, Should and Consider – and those that are enforced through national policy or initiatives are marked accordingly.

At a minimum, producers must (1) identify areas at risk of erosion, compaction and leaching; (2) create a policy related to organic matter management; (3) implement a soil analysis program; (4) assess field conditions prior to cultivation; (5) create a nutrient management plan; (6) calculate nitrogen needs to limit leaching risk; and (7) record all nutrient management applications. In addition, the Framework encourages producers to (1) maintain a soil map that identifies the main soil types and associated characteristics on the farm; (2) plan crop rotations a minimum of three years in advance; (3) establish a Soil Management Plan; (4) keep current with technical information and advice related to soil management; (5) maintain soil cover during non-production periods; (6) take measures to prevent soil compaction, improve soil structure and increase porosity; (7) monitor soil pH; and (8) record all soil operations by type of crop and/or field. The ultimate objective of the soil guidelines is to demonstrate that conservation and improvement of soil resources is essential to Integrated Farming.

4.4 European Union Organic Production and Labelling of Organic Products

The European Union's Organic Regulation[33] establishes the legal framework for all levels of production, distribution, control and labelling of organic products for purchase and trade in the EU. The regulations apply to living or unprocessed products, processed foods, animal feed and seed, and propagating material.

In terms of soil parameters, the regulations generally state that organic plant production should maintain and enhance soil fertility, prevent soil erosion and derive nutrients primarily from the soil. More specifically, the regulations identify soil fertility management, crop rotation, appropriate choice of crop species and varieties, recycling of organic materials and judicious use of fertilisers, soil conditions and plant protection products as "essential" elements of an organic plant production system. The crucial role of soils in organic production systems is succinctly captured in the only soil-related principle explicitly included in the EU regulations, which states that "organic farming shall be based on the maintenance and enhancement of soil life and natural soil fertility, soil stability and soil biodiversity, preventing and combating soil compaction and soil erosion, and the nourishing of plants primarily through the soil ecosystem."

4.5 Generic Fairtrade Standards for Small Producers' Organisations

The Fairtrade Labelling Organisation International (FLO) Standards for Small Producers' Organisations[34] were created to help small producers in the Global South overcome barriers to economic development and empowerment. In order to qualify for certification, small farmers must comply with criteria across several impact categories that address issues related to social development, socioeconomic development, environmental development and labour conditions.

The primary objective of the standards relative to soil is for producers to maintain and enhance soil fertility and structure. Minimum requirements toward this goal include taking measures to reduce and/or prevent soil erosion, establishing on-farm guidelines based on successful techniques and practices designed to ensure soil fertility and improved soil structure, and evaluating overall compliance to the standard in order to identify areas of improvement. Additional credit is awarded for Progress Requirements, namely measures taken to ensure that water management, tillage practices and/or the use of irrigation water does not lead to contamination, desertification or excessive salinization of soil.

4.6 GlobalG.A.P. Standards

GLOBALG.A.P. is private sector body that sets voluntary standards for certification of agricultural products and processes across the world. The standards[35] are intended to serve as a "global reference system" for other sustainability standards and to assure purchasers about the sustainability attributes of how food is produced. GlobalG.A.P. is a business-to-business label that is not directly visible to consumers.

The GlobalG.A.P. standards offer producers a single, integrated approach that includes baseline requirements for all participating farms, along with modular applications for different agricultural product groups. Soil parameters are covered under the All Farm Base, Crops Base, Fruit and Vegetables module and Plant Propagation module. The All Farm Base[36] asks producers to conduct a risk assessment of the production site undergoing certification. From a soils perspective, the assessment must address soil type, structural stability for intended use, susceptibility to erosion, chemical suitability for intended crops and drainage patterns. The Crops Base[37] requires producers to adopt techniques that improve or maintain soil structure, avoid soil compaction and reduce the possibility of soil erosion. Preparation of soil maps is recommended but not essential for certification. Under the Fruits and Vegetables module,[38] producers that use soil fumigants must provide a written justification for doing so. No other soil management practices are addressed. The Plant Propagation[39] module includes all of the soil parameters covered above, the only difference being that implementing practices to improve or maintain soil structure and avoid soil compaction are recommended but not required. Overall, the objective of the soil component of GlobalG.A.P is to demonstrate that soil conservation and improvement is essential to long-term soil fertility, sustained yields and profitability of the farming operation.

4.7 IFOAM Norms for Organic Production and Processing

The IFOAM Norms,[40] developed by the International Federation of Organic Agriculture Movements (IFOAM), provide a common set of international standards for organic production and processing, create a common system of verification and market identity, and facilitate the trade of organic products. The standards address the specific principles, recommendations and required baseline practices that guide organic crop production and ensure organic integrity in the handling and processing of organic commodities. Accredited IFOAM certification bodies can certify producers that meet the requirements of the standard.

The IFOAM norms cover seven general components of organic agricultural production: Organic Ecosystems, Crop Production, Animal Husbandry, Processing and Handling, Labelling, Social Justice and Aquaculture Production. Soil conservation, included under Organic Ecosystems, is addressed through a series of recommendations and required practices. To meet the baseline, producers must take appropriate measures to prevent soil erosion, restrict vegetation burning for purposes of land preparation to a minimum, ensure that crop production practices return nutrients, organic matter and other resources to the soil, and take relevant action to prevent or remedy soil salinisation. Additional recommendations include minimising topsoil erosion, preventing soil compaction and mitigating soil degradation through any range of practices deemed appropriate by the producer.

4.8 LEAF Marque Global Standard

The LEAF (Linking Environment and Farming) Marque standard[41] is global in scope and aims to give consumers confidence in the quality and environmentally friendly attributes of food and other agricultural products. Producers must conduct a self-assessment of their farm and farm-management practices across seven categories: Organisation and Planning, Soil Management and Crop Nutrition, Crop Protection, Pollution Control and Waste Management, Energy and Water Efficiency, Wildlife and Landscape, and Animal Husbandry and the Environment. In order to qualify for certification, producers must earn a minimum number of points within each category.

Required practices under Soil Management and Crop Nutrition include: (1) consulting a qualified agronomist for crop nutrition advice; (2) following a nutrient management plan; (3) using soil-mapping techniques to identify areas prone to compaction, erosion, run-off and leaching; (4) conserving and building soil organic matter; (5) monitoring soils and crops prone to trace element deficiencies; (6) implementing a long-term cropping and rotation plan; (7) carrying out field operations under appropriate conditions to protect soil structure; (8) maintaining records of fertiliser applications; and (9) applying soil nutrients correctly. Recommended practices covered by this section include: (1) estimating soil nitrogen supply available to plants; (2) measuring nitrogen efficiency per unit production; and (3) recording all cultivations and field operations.

4.9 RTRS Standard for Responsible Soy Production

The Roundtable on Responsible Soy (RTRS) Standard for Responsible Soy Production[42] emerged out of a multi-stakeholder dialogue to promote economically viable, socially equitable and environmental sound soy production across the world. The resulting certification scheme is applicable at a global level, is relevant to soy production for animal feed, human consumption and biofuels, is appropriate for producers of all scales and types and meets global sustainability goals to assure access to a wide range of markets.

The core principles of the RTRS standard cover Legal Compliance and Good Business Practice, Responsible Labour Conditions, Responsible Community Relations, Environmental Responsibility and Good Agricultural Practices. Soil parameters addressed include soil carbon, soil erosion, soil quality and soil organic matter. Under the Environmental Responsibility provisions, producers must quantify changes in soil carbon levels and take appropriate measures to ameliorate negative trends. Requirements addressed under Good Agricultural Practices include minimising and controlling soil erosion, maintaining or improving soil quality and monitoring soil organic matter.

4.10 RSB Principles and Criteria for Sustainable Biofuel Production

The Roundtable on Sustainable Biofuels (RSB) is an international, multi-stakeholder effort aimed at ensuring the sustainability of biofuels production and processing. To meet this objective, RSB has developed a set of certification guidelines[43] for the production and processing of biofuel feedstock and raw material that is applicable to biofuel operations along the entire supply chain. Producers first self-assess their performance across twelve principle areas identified within the RSB Principles and Criteria for Sustainable Biofuel Production before submitting their certification application. These principle areas include: Legality; Planning, Monitoring and Continuous Improvement; Greenhouse Gas Emissions; Human and Labor Rights; Rural and Social Development; Local Food Security; Conservation; Soil; Water; Air; Use of Technology, Inputs, and Management of Waste; and Land Rights.

The objective of the RSB soil principle is to reverse soil degradation related to biofuels production and to maintain soil health. To that end, producers must implement a soil management plan designed to maintain or enhance soil physical, chemical and biological characteristics. At a minimum, the management plan must include provisions to: (1) minimise soil erosion through use of sustainable practices such as crop rotation, vegetative ground cover and terracing; (2) avoid the use of agrochemicals prohibited by the standard; and (3) promote long-term soil stability and organic matter content. Progress requirements include improving soil health through direct planting, permanent soil cover, crop rotation, fallow areas with natural or planted vegetation and other soil conservation practices.

4.11 RSPO Principles and Criteria for Sustainable Palm Oil Production

The Principles and Criteria for Sustainable Palm Oil Production,[44] developed by the Roundtable on Sustainable Palm Oil Production (RSPO), is aimed at advancing economically viable, environmentally appropriate, and socially beneficial palm oil operations. Each of eight principle areas contain specific sustainability indicators that producers must demonstrate in order to qualify for certification. These principle areas include: Commitment to Transparency; Compliance with Applicable Laws and Regulations; Commitment to Long-Term Economic and Financial Viability; Use of Appropriate Best Practices; Environmental Responsibility and Conservation of Natural Resources and Biodiversity; Responsible Consideration of Employees, Individuals and Communities; Responsible Development of New Plantings; and Commitment to Continuous Improvement.

Criteria related to soils address fertility, erosion and suitability for production. Producers are required to (1) implement practices that maintain and/or improve soil fertility to levels that ensure optimal and sustained yield; (2) minimise and control erosion and degradation of soils; (3) avoid extensive planting on marginal and fragile soils; and (4) use soil surveys for production management and planning. Relevant indicators that producers may monitor to determine whether they are meeting these criteria include: soil organic matter, net fertiliser inputs, plant nutrient status, soil structure, percent of planting on slopes above a soil-specific limit, percent of ground surface protected from raindrop impact, plant rooting depth, soil moisture and long-term palm oil yields.

4.12 SAN Sustainable Agriculture Standard

The Sustainable Agriculture Network (SAN) is coalition of non-profit, conservation-oriented organisations that develop standards to promote the social and environmental sustainability of agriculture. The SAN Sustainable Agriculture Standard,[45] originally developed to encourage the implementation of best management practices in Latin American agriculture, is followed by tens of thousands of producers across the world who grow such crops as coffee, tea, banana, citrus, cocoa, vanilla, macadamia, rubber, soy, palm oil and sugarcane. In order to qualify for certification, producers must comply with a minimum of 80% of the standard's total criteria across ten principle areas: Social and Environmental Management, Ecosystem Conservation, Wildlife Protection, Fair Treatment and Good Working Conditions for Workers, Occupational Health and Safety, Community Relations, Integrated

Table 3 Summary of standard tools relating soils and sustainable agriculture.

Standard	*Crop(s)*	*Soil properties*	*Assessment method(s)*
Basel Criteria for Responsible Soy Production	Soybean	Soil type, soil nutrients, soil erosion potential, soil structure	Soil maps/surveys, producer observation
Better Cotton Initiative	Cotton	Soil structure, soil organic matter, soil nutrients, soil erosion potential	Soil and foliage tests, producer observation
European Integrated Farming Framework	Crops and livestock	Soil organic matter, soil structure, soil nutrients, soil erosion potential, soil leaching potential, soil type, soil permeability, soil pH	Soil maps/surveys, soil and foliage tests, producer observation
EU Organic Production and Labelling of Organic Products	Crops and livestock	Soil nutrients, soil erosion potential, soil biological activity, soil structure	None specified
Generic Fairtrade Standards for Small Producers' Organizations	Crops and agroforestry	Soil nutrients, soil structure, soil erosion potential, soil salinity	Producer observation
GlobalG.A.P. Standards	Crops, livestock, aquaculture	Soil structure, soil type, soil erosion potential, soil nutrients, soil drainage	Soil maps/surveys, soil and foliage tests, producer observation
IFOAM Norms for Organic Production and Processing	Crops and livestock	Soil erosion potential, soil nutrients, soil organic matter, soil salinization, soil structure	None specified
LEAF Marque Global Standard	Agronomic and horticultural crops	Soil type, soil nutrients, soil structure, soil erosion potential, soil leaching potential, soil organic matter	Soil maps/surveys, soil and foliage tests, producer observation
RTRS Standard for Responsible Soy Production	Soybeans	Soil carbon, soil erosion potential, soil organic matter	Soil and foliage tests, producer observation
RSB Principles and Criteria for Sustainable Biofuel Production	Biomass crops	Soil structure, soil organic matter; overall physical, chemical and biological properties	Soil maps/surveys, soil and foliage tests, producer observation
RSPO Principles and Criteria for Sustainable Palm Oil Production	Palm oil	Soil nutrients, soil erosion potential, soil type, soil organic matter, soil structure, soil moisture	Soil maps/surveys, soil and foliage tests, producer observation
SAN Sustainable Agriculture Standard	Coffee, cocoa, banana, tea, pineapple, flowers, foliage, citrus, aloe vera, apple, avocado, cherry, grapes, heart of palm, kiwi, macadamia, mango, pear, rubber, vanilla	Soil erosion potential, soil nutrients, soil structure, soil organic matter, soil type, soil carbon, soil moisture holding capacity, soil permeability	Soil and foliage tests

Crop Management, Soil Management and Conservation, and Integrated Waste Management. Authorised certifiers verify conformance to the standard.

The objective of the Soil Management and Conservation principle is to promote the long-term improvement of soils for sustained agricultural production. To that end, participating producers must meet the following requirements: (1) follow a soil erosion prevention and control program based on soil properties and characteristics, climatic conditions and utilised crop production practices; (2) implement a soil and crop fertilisation program based on soil properties and characteristics, soil and foliage sampling and analysis, and advice from an agricultural professional; (3) use and expand vegetative ground cover to reduce erosion and improve soil fertility, structure and organic matter content; (4) promote the use of fallow areas with natural or planted vegetation with the aim of restoring natural fertility; and (5) locate new production areas only on land suitable for the intensity of agricultural production planned. In addition, the SAN Climate Module, which attempts to mitigate the impact of agriculture on climate change, requires producers to maintain or increase soil carbon stocks by incorporating practices that reduce tillage, recycle crop nutrients and residues and optimise the soil water's retention and infiltration. Table 3 summarises the 12 global standards that are currently being used and developed.

5 Discussion and Conclusions

Assessing sustainable land management practices and their implications for soil with consistency and credibility, and reflecting insights from agricultural and environmental science and economics, are as difficult as defining sustainability. Some of the problems are the spatial and temporal borders that need to be chosen for its assessment[46,47] and the selection of indicators and standards to evaluate sustainability in a given locality.[48,49] Long-term data are imperative to evaluate unidimensional trends and important interactions that affect the sustainability of land-management practices, but they are scarce for many parts of the world.

Sustainability, although a dynamic concept, implies some sort of equilibrium or steady state.[50] Indicators, defined as attributes that measure or reflect conditions of sustainability,[48] should therefore not show a significantly declining trend.[51] Many soil properties significantly change with time, and it can be argued that land-use systems in which soil changes take place are not sustainable in the long-term. Such a conclusion demands a rigorous assessment of soil change and a set of soil properties that can be used as indicators. Each soil property shows a degree of natural variation that is affected, for example, by soil management and the overall cropping system. There is a cost to collect such data, and, for general assessment, there is a trade-off between additional costs and the extra information obtained. In the previous sections, various sustainability standards and reporting protocols across the USA and around the world have been reviewed to demonstrate the possible range of sustainability indicators, which could be physical, chemical, biological, economic, social or other. The identification of suitable indictors for soil is complicated by the multiplicity of physical, chemical and biological factors that control biogeochemical processes and their variation in intensity over time and space.[52] A set of basic indicators of soil changes has not been defined, largely due to the range across which soil properties vary in magnitude and importance. Pedotransfer functions can be used to estimate soil attributes that are lacking; for example, the CEC can be estimated from the clay content and type, and the organic C content.

Agricultural decision-making will not and does not need to await the rigorous development and endorsement of a full framework that incorporates all of the determinants of sustainability, such as standards, indicators, criteria and threshold values. Analyses often reveal that some, but not all, sustainability requirements are met and hence a system can be evaluated as "partially" or "conditionally" sustainable. Being able to track the general process towards the goal may be more useful than setting specific, rigid targets to be achieved.[53] This seems a useful alternative to the Framework for the Evaluation of Sustainable Land Management (FESLM) approach. There is a whole range of new observational and measurement techniques in soil science[54,55] that will greatly facilitate the evaluation and reporting of soils in sustainable agriculture and food systems.

In discussions of agricultural sustainability, it is our contention that we are at a critical inflection point with respect to the objective approaches to describe the consequences and implications of agricultural practices on soil. We have reviewed standards where Tier 1 sustainability is measuring parameters of importance in diverse, robust, resilient agricultural systems that adequately provide food for future human populations. In general, these approaches still fall short of moving agriculture from a fundamentally extractive process to one that is fundamentally regenerative and restorative for soil resources. Tier 2 sustainability requires, however, a commitment to reduce the environmental footprint and improve outcomes in human dimensions. At present, many businesses and organisations have begun to use "corporate sustainability", "green strategies" and the concept of "continuous improvement" and therefore are transitioning into this second tier of sustainability. There is concern that performance of agricultural systems is profoundly discontinuous and, in the face of climate change, shows more variability, especially across dimensions of time and space. As such, we propose the concept of Tier 3 sustainability, whereby we commit to actions, recognising local-to-global continua that will move us, in aggregate, into planetary safe space with respect to outcomes in human, environmental and ecological dimensions.[56] In this tier, the commitment is to move toward agricultural systems that are fundamentally regenerative, consistent with practices that could be used over millennia with steady or improving productivity.

In this chapter, we have summarised some of the current approaches in the USA and global sustainable agriculture regarding soil attributes and practices.

These are used in either a Tier 1 mode, where the objective is simply to measure or observe, or a Tier 2 mode where the objective is to measure in order to improve. We conclude with a call to action aimed to achieve and test our progress toward agriculture in long-term balance with our resources – an agriculture that improves soils in every dimension and protects against soil degradation, which is a growing threat in the face of increasingly frequent extreme weather in the Anthropocene era. But the present is not reassuring. Each year it is estimated that 12 000 million ha of land degrade to the point where the land is lost to agriculture, equivalent to 20 million tonnes of grain.[57,58] Beyond science, there is an urgent need to understand and affect the decisions we make about soil resources for our long-term survival and quality of life and for the integrity of the planetary system.

References

1. A. E. Hartemink and A. McBratney, *Geoderma*, 2008, **148**(2), 123–129.
2. UNEP, *Global Environmental Outlook GEO4 - Environment for Development*,United Nations Environment Programme, Nairobi, 2007.
3. P. A. Sanchez, S. Ahamed, F. Carre, A. E. Hartemink, J. Hempel, J. Huising, P. Lagacherie, A. B. McBratney, N. J. McKenzie, M. D. Mendonca-Santos, B. Minasny, L. Montanarella, P. Okoth, C. A. Palm, J. D. Sachs, K. D. Shepherd, T. G. Vagen, B. Vanlauwe, M. G. Walsh, L. A. Winowiecki and G. L. Zhang, *Science*, 2009, **325**(5941), 680–681.
4. A. Marchetti, C. Piccini, R. Francaviglia and L. Mabit, *Pedosphere*, 2012, **22**(2), 230–242.
5. E. L. Thomaz and J. C. Luiz, *Land Degradat. Dev.*, 2012, **23**(1), 72–81.
6. F. Khormali and M. Ajami, *Geoderma*, 2011, **167–68**, 274–283.
7. R. Lal, *Climate Change and Food Security in South Asia*, Springer, 2011,137–152.
8. Y. A. Phillis and L. A. Andriantiatsaholiniaina, *Ecol. Econom.*, 2001, **37**(3), 435–456.
9. D. J. Greenland, in *Soil Science and Sustainable Land Management in the Tropics*, ed. J. K. Syers and D. L. Rimmer, CAB International, Wallingford, 1994, pp. 1–15.
10. A. Young, *Agroforestry for Soil Management*, 2nd edn, 1997, CAB International, Wallingford, 1997.
11. J. Bouma, *Soil Sci. Soc. Am. J.*, 1994, **58**, 645–646.
12. M. C. Scholes, M. J. Swift, O. W. Heal, P. A. Sanchez, J. S. I. Ingram and R. Dalal, in *The Biological Management of Tropical Soil Fertility*, ed. P. L. Woomer and M. J. Swift, John Wiley & Sons, Chichester, 1994, pp. 1–14.
13. M. J. Swift, in *Soil Resilience and Sustainable Land Use*, ed. D. J. Greenland and I. Szabolcs, CAB International, Wallingford, 1994, pp. 235–248.
14. A. E. Hartemink, *Soil Use and Management*, 2008, **24**(4), 327–330.

15. The Keystone Center, *Field to Market Fieldprint Calculator*; http://www.fieldtomarket.org/calculator.php, 2012, accessed 24/04/2012.
16. Stewardship Index for Specialty Crops, and Ag Innovations Network, *User's Guide to SISC Farm Level Metrics and Calculator Beta Version 1.1*; http://www.stewardshipindex.org/docs/Guide to SISC Metrics.pdf, 2011, accessed 10/05/2012.
17. S. S. Andrews, D. L. Karlen and C. A. Cambardella, *Soil Sci. Soc. Am. J.*, 2004, **68**(6), 1945–1962.
18. Sustainable Food Lab, University of Aberdeen, and Unilever, *The Cool Farm Tool, a User's Guide*; http://www.unilever.com/images/The Cool Farm Tool - A User's Guide_tcm13-263218.pdf, 2011, accessed 02/05/2012.
19. D. Sonke, A. Arnold, G. Ludwig and J. Dlott. *California Almond Board Irrigation Management Module*; http://www.almondboard.com/Growers/Sustainability/SustainabilityModules/Pages/Default.aspx, 2010, accessed 02/05/2012.
20. D. Sonke, A. Arnold, G. Ludwig and J. Dlott. *California Almond Board Nutrient Management Module*; http://www.almondboard.com/Growers/Sustainability/SustainabilityModules/Pages/Default.aspx, 2010, accessed 02/05/2012.
21. Council on Sustainable Biomass Production, *Draft Provisional Standard for Sustainable Production of Agricultural Biomass;* http://www.csbp.org/files/survey/CSBP_Provisional_Standard.pdf, 2010, accessed 02/05/2012.
22. Demeter Association, *Demeter Biodynamic Farm Standard*; http://www.demeter-usa.org/downloads/Demeter-Farm-Standard.pdf, 2010, accessed 02/05/2012.
23. Food Alliance, *Food Alliance Whole Farm/Ranch Inspection Tool*; http://foodalliance.org/certification/producer/WholeFarm.pdf, 2008, accessed 02/05/2012.
24. University of Wisconsin, Wisconsin Potato and Vegetable Growers Association, International Crane Foundation, and World Wildlife Foundation, *Wisconsin Vegetable Sustainability Standards*, 2011.
25. Lodi Winegrape Commission, *The Lodi Rules For Sustainable Winegrowing: Farming Practices Standards;* http://www.lodiwine.com/Lodi_Rules_Farming_Practices.pdf, 2009, accessed 02/05/2012.
26. United States Department of Agriculture, *National Organic Program*; http://www.ams.usda.gov/AMSv1.0/nop, accessed 21/04/2012.
27. United States Department of Agriculture, *National Organic Program*, Part 205.203, Soil Fertility and Crop Nutrient Management Practice Standard, Organic Foods Production Act; http://ecfr.gpoaccess.gov/cgi/t/text/text-idx?c=ecfr&sid=139c97f3d58c0f68e32e8f52bed5f83c&rgn=div8&view=text&node=7:3.1.1.9.32.3.354.4&idno=7, accessed 21/04/2012.
28. United States Department of Agriculture, *National Organic Program*, Part 205.205, Crop Rotation Practice Standard, Organic Foods Production Act; http://ecfr.gpoaccess.gov/cgi/t/text/text-idx?c=ecfr&sid=

139c97f3d58c0f68e32e8f52bed5f83c&rgn=div8&view=text&node=7:3.1.1.9.32.3.354.6&idno=7, accessed 21/04/2012.

29. Central Coast Vineyard Team, *Sustainability in Practice (SIP) Vineyard Certification Program*; http://www.vineyardteam.org/files/sip/SIP Vineyard Standards 2012.pdf, 2012, accessed 02/05/2012.
30. ProForest for Coop Switzerland and WWF Switzerland, *The Basel Criteria for Responsible Soy Production*; http://www.proforest.net/objects/publications/basel-criteria-august-04.pdf, 2004, accessed 02/05/2012.
31. Better Cotton Initiative, *Benchmarks for Large Farm Employers;* http://www.bettercotton.org/index/140/better_cotton_system.html, 2010, accessed 02/05/2012.
32. European Initiative for Sustainable Development in Agriculture, *European Integrated Farming Framework*; http://www.sustainable-agriculture.org/stuff/EISA-Framework-english-040810.pdf, 2010, accessed 02/05/2012.
33. Council of the European Union, *Council Regulation on Organic production and Labelling of Organic Products and Repealing Regulation (EEC)*; http://eur-lex.europa.eu/LexUriServ/LexUriServ.do?uri=OJ:L:2007:189:0001:0023:EN:PDF), 2007, accessed 30/04/2012.
34. Fairtrade Labelling Organization International, *Generic Fairtrade Standards for Small Producers' Organisations*; Bonn, Germany 2009.
35. GLOBALG.A.P. Standards; http://www.globalgap.org/cms/front_content.php?idcat=3, 2012, accessed 02/05/2012.
36. GLOBALG.A.P, *Control Points and Compliance Criteria – All Farm Base*; http://www.globalgap.org/cms/upload/The_Standard/IFA/Version_4.0-1_Feb2012/English/CPCC/120206_gg_ifa_cpcc_af_eng_v4_0-1.pdf, 2011, accessed 03/05/2012.
37. GLOBALG.A.P, *Control Points and Compliance Criteria – Crops Base*; http://www.globalgap.org/cms/upload/The_Standard/IFA/Version_4.0-1_Feb2012/English/CPCC/120206_gg_ifa_cpcc_cb_eng_v4_0-1.pdf, 2011, accessed 03/05/2012.
38. GLOBALG.A.P, *Control Points and Compliance Criteria – Fruits and Vegetables*; http://www.globalgap.org/cms/upload/The_Standard/IFA/Version_4.0-1_Feb2012/English/CPCC/120206_gg_ifa_cpcc_fv_eng_v4_0-1.pdf, 2011, accessed 03/05/2012.
39. GLOBALG.A.P, *Control Points and Compliance Criteria – Plant Propagation Material*; http://www.globalgap.org/cms/upload/The_Standard/Plant_Propagation_Material/English/CPCC/GG_EG_PPM_CPCC_ENG_V2_0_Mar08.pdf, 2008, accessed 03/05/2012.
40. IFOAM, *The IFOAM Norms for Organic Production and Processing*; http://www.ifoam.org/about_ifoam/standards/norms/norm_documents_library/Norms_ENG_V4_20090113.pdf, 2005, accessed 01/05/2012.
41. LEAF Marque, *LEAF Marque Global Standard*; http://www.leafuk.org/resources/000/557/691/LEAF_Marque_Global_standard_ver_9.0_2010.pdf, 2012, accessed 02/05/2012.

42. Roundtable on Responsible Soy Association, *RTRS Standard for Responsible Soy Production*; http://www.responsiblesoy.org/index.php?option=com_docman&task=doc_download&gid=290&Itemid=19&lang=en, 2010, accessed 04/05/2012.
43. Roundtable on Sustainable Biofuels, *RSB Principles & Criteria for Sustainable Biofuel Production*; http://rsb.epfl.ch/files/content/sites/rsb2/files/Biofuels/Version 2/PCs V2/11-03-08 RSB PCs Version 2.pdf, 2010, accessed 04/05/2012.
44. Roundtable on Sustainable Palm Oil, *RSPO Principles and Criteria for Sustainable Palm Oil Production*; http://www.rspo.org/sites/default/files/RSPO Principles & Criteria_0.pdf, 2007, accessed 04/05/2012.
45. Sustainable Agriculture Network, *Sustainable Agriculture Standard*, 2010.
46. L. O. Fresco and S. B. Kroonenberg, *Land Use Policy*, 1992, **9**, 155–168.
47. G. K. Heilig, *Int. J. Sustainable Dev. World Ecol.*, 1997, **4**(1), 1–16.
48. A. J. Smyth and J. Dumanski, *Can. J. Soil Sci.*, 1995, **75**, 401–406.
49. C. Pieri, J. Dumanski, A. Hamblin and A. Young, 1995, Land quality indicators. World Bank Discussion Paper 315, Washington DC.
50. J. R. O'Callaghan and G. C. L. Wyseure, in *Soil Science and Sustainable Land Management in the Tropics*, ed. J. K. Syers and D. L. Rimmer, CAB International, Wallingford, 1994, pp. 27–39.
51. W. E. Larson and F. J. Pierce, in *Defining Soil Quality for a Sustainable Management*, ed. J. W. Doran, D. C. Coleman, D. F. Bezdicek and B. A. Stewart, SSSA-ASA, Madison, 1994, pp. 37–51.
52. J. W. Doran and T. B. Parkin, in *Methods for Assessing Soil Quality*, ed. J. W. Doran and A. J. Jones, SSSA, Madison, 1996, pp. 25–37.
53. S. A. El-Swaify, *J. Soil Water Conserv.*, 2000, **55**, 2–3.
54. A. B. McBratney, B. Minasny and R. Viscarra Rossel, *Geoderma*, 2006, **136**(1–2), 272–278.
55. R. Viscarra Rossel, V. I. Adamchuk, K. A. Sudduth, N. J. McKenzie and C. Lobsey, Adv. Agron., 2011, **113**, 237–282.
56. J. Beddington, M. Asaduzzaman, M. Clark, A. Fernandez, M. Guillou, M. Jahn, L. Erda, T. Mamo, N. Van Bo, C.A. Nobre, R. Scholes, R. Sharma and J. Wakhungu, *Achieving Food Security in the Face of Climate Change: Final Report from the Commission on Sustainable Agriculture and Climate Change*, CGIAR Research Program on Climate Change, Agriculture and Food Security (CCAFS), Copenhagen, Denmark, 2012.
57. Z. G. Bai and D. Dent, *Ambio*, 2009, **38**(3), 150–156.
58. Z. G. Bai, D. L. Dent, L. Olsson and M. E. Schaepman, *Soil Use Manage.*, 2008, **24**(3), 223–234.

Agrobiodiversity and Potential Use for Enhancing Soil Health in Tropical Soils of Africa

FREDRICK O. AYUKE*[a], NANCY K. KARANJA[a], JULIUS J. OKELLO[b], PETER M. WACHIRA[c], GERALD K. MUTUA[d], DAVID K. LELEI[e] AND CHARLES K. K. GACHENE[a]

[a] Department of Land Resource Management and Agricultural Technology, University of Nairobi, P.O Box 29053-00625, Nairobi, Kenya; [b] Department of Agricultural Economics, University of Nairobi, P.O Box 30197-00100, Nairobi, Kenya; [c] School of Biological Sciences, University of Nairobi, P.O Box 30197-00100, Nairobi, Kenya; [d] Department of Plant Sciences, University of Nairobi, P.O Box 30197-00100, Nairobi, Kenya; [e] World Agroforestry Centre, P.O Box 30677, Nairobi, Kenya*Corresponding author
*E-mail: fredrick.ayuke@yahoo.com

ABSTRACT

Land degradation and soil fertility decline are often cited as major constraints to crop production in sub-Saharan Africa (SSA). As mineral and organic fertilisers are often limited in quantity and quality, soil fertility research has focused on developing integrated management strategies to address soil fertility decline. Soil biota are an essential component of soil health and constitute a major fraction of global terrestrial biodiversity. Within the context of Integrated Soil Fertility

Issues in Environmental Science and Technology No. 35
Soils and Food Security
Edited by R.E. Hester and R.M. Harrison

Published by the Royal Society of Chemistry, www.rsc.org

Management (ISFM), soil biota are responsible for the key ecosystem functions of decomposition and nutrient cycling, soil organic matter synthesis and mineralisation, soil structural modification and aggregate stabilisation, nitrogen fixation, nutrient acquisition, regulation of atmospheric composition, the production of plant growth substances and the biological control of soil-borne pests and diseases. Soil biological processes are not as well understood as are soil physical and chemical properties, creating opportunities for breakthroughs in biotic function to provide better services to agriculture. These services accrue through two basic approaches: indirectly, as a result of promoting beneficial soil biological processes and ecosystem services through land management, or directly, through the introduction of beneficial organisms to the soil. Because of their sensitivity to disturbance and their importance in redistributing and transforming organic inputs, some of the soil biota groups, such as earthworms and termites, represent an important indicator of soil quality. In this chapter we have highlighted the importance of soil biodiversity, especially its potential use for enhancing soil health in tropical soils of SSA.

1 Introduction

Lack of food is of central concern in Africa and presents a fundamental challenge for human welfare and economic growth. Increased population growth coupled with limited resources in many developing countries has contributed to increased levels of poverty, resulting in land sub-division and environmental and land degradation.[1] The net result is small farms, low production and increasing landlessness.[1,2]

Land degradation and soil infertility or nutrient depletion are therefore considered as major threats to food security and natural conservation in sub-Saharan Africa (SSA). Increasing population pressures and widespread food deficits in SSA have compelled national programmes and international donors to place a high priority on increased agricultural productivity and alleviation of poverty among the small-scale farmers. Despite this, few new technical packages capable of increasing net returns without deteriorating the environment have been developed. As such, the challenge of increasing crop yields to sustain the growing population is persistent.

2 Description of Soils in Sub-Saharan Africa

The soils pattern in the SSA countries is intricate because of large differences in altitude, topography, geology and climate. In particular, they are based on a wide range of parent materials, ranging from sedimentary, metamorphic to volcanic rocks. This has resulted in the formation of different soil profiles with varying texture (which in most cases determines the ability of the soil to hold and release moisture), depth and inherent soil fertility. Most of the tropical soils have serious constraints to crop production; among them, extended periods / seasonal moisture stress, perhaps the overriding constraint in much of African soils (about 14% of Africa is relatively free of moisture stress), salinity, sodicity, acidity, drainage, shallow rooting depth and fertility problems. For agricultural planning, it is therefore essential that the distribution, extent, limitations and potential of these different soil types be appreciated. The general occurrence, characteristics, use and management of the major soils found in Sub-Saharan Africa are summarised in Table 1.

Tables 2 and 3 show the physical and chemical characteristics of the major soil types of Kenya and for some west African countries (Liberia, Nigeria, Ghana, Togo, Burkina Faso, Benin, Niger, Mali and Gambia). The two tables show that most of the soils are dominated by Lixisols, Acrisols, Luvisols, Nitisols, Alisols and Ferralsols. In particular, the Kenyan soils represent the major soil types occurring in east and central Africa regions. Table 2 shows soil physical and chemical properties for 45 fertiliser trails in the high and medium potential areas of Kenya for a wide range of major soil types found in the Kenyan highlands. The soils were selected for fertiliser recommendation for different agroecological zones in Kenya. It is noted in Table 2 that even within the same soil group, the soil properties can vary greatly. For instance, soils for Mumias and Chepkumia are both Acrisols, yet the texture, organic carbon and total nitrogen vary greatly in both sites (sandy loam texture, 5 g kg^{-1} organic carbon (OC) and 0.6 g kg^{-1} N for Mumias and clay loam texture, 28 g kg^{-1} OC and 4.5 g kg^{-1} N). Such variations in soil properties do re-emphasise the need for specific fertiliser recommendations rather than blanket recommendations as is mostly the case in Sub-Saharan Africa. Most of these tropical soils have undergone ferrugination and ferralitisation processes, an indication of soils that have undergone intense chemical weathering. As a result, these soils are of low inherent fertility. Coupled with low fertiliser inputs, on-farm nutrient balance is, in most cases, negative. For instance, nutrient balance calculations revealed that annual nutrient depletion in Kisii (Kenya) was 112, 2.5 and 70 kg ha^{-1} for N, P and potassium (K), respectively,[3] whereas in southern Mali the values were 25, 0 and 20 kg ha^{-1} for N, P and K, respectively.[4] In Kisii, removal of nutrients in the harvested product was the strongest contributor to the negative balance, followed by run-off and erosion. Work carried out in Kenya on the effect of erosion on soil fertility supports these findings.[5] Changes in soil pH (regression coefficient, r = 0.77), OC (r = 0.59) and total nitrogen (TN) (r = 0.71) were highly and positively correlated with soil loss, while maize grain yield was highly and negatively correlated with soil loss (r = −0.91). In the same study, sediment from the plots was 247% to 936% richer in P than the soil from which it originated. The data indicate that nutrient loss due to erosion is one of the major causes of soil fertility depletion of Sub-Saharan African soils. Soil degradation of arable land, through loss of soil organic matter (SOM) and soil structural stability also results from soil tillage and the removal of plant biomass. In many tropical cropping systems, little or no agricultural residues are returned to the soil as these are either burnt to clear the ground for crop planting, utilised as fuel, or grazed by livestock.[6] The loss of SOM and the associated deterioration of soil physical, chemical and biological fertility associated with continuous cropping and sub-optimal fertiliser use frequently result in a decline in biomass productivity and crop yields and present great challenges to many farmers in Sub-Saharan Africa.[7]

Of great concern are the low levels of phosphorus for the majority of SSA soils. For instance, of the 147 soil samples analysed for P in four irrigation projects in Rwanda, only 10 had adequate levels of P. The soils were predominantly orthic Ferralsols with their integrades, namely, ferralo-orthic Luvisols and Lixisols. Exchangeable acidity was on average >1.0 meq. In a study investigating the relationships between phosphorus sorption index (PSI) and selected soil chemical properties of these soils in Rwanda, it was found that the pH of these soils was variable, ranging from 5.3 to 5.6, *i.e.* moderately to strongly acidic. The PSI for the soils ranged from 25.93 to 295.52 ppm. The wide range of the differences in PSI indicates that blanket phosphate recommendations may not be a good strategy for most of the soils in the Sub-Saharan African countries as it may lead to under or over application of P.

3 Land Degradation in Cropping Systems

The recognised form of land degradation affecting the major soil types in sub-Saharan Africa are erosion, physical and chemical degradation, which includes salinisation, sodification, acidification and the depletion of plant nutrient content in the soil. Biological degradation is also a major contributor, leading to loss of soil organic matter and soil biodiversity. All these forms of degradation lead to a lowering of soil fertility and land productivity.[8] Land degradation is now recognised as being one of the major contributors to the persistent food deficit and high poverty levels in Sub-Saharan Africa. According to Gachene and Kimaru,[9] concerted and well-planned action needs to be taken to build soil fertility and minimise land degradation on small-scale farms. Some of the important action points are developing well-defined and specific activities to enhance plant nutrient levels as a long-term programme through consistent use of both organic and inorganic fertilisers. According to World Bank figures, Africa uses only 14 kg of fertiliser per hectare compared with 1150–200 kg in East Asia and Europe. Use of both organic and inorganic fertilisers has resulted in improved soil physical and chemical properties and increased crop yields for some of the highly weathered tropical soils,[10,11] giving adequate attention to the problem of soil acidity and finding better ways of promoting plant nutrient availability and uptake,[12] developing and adapting suitable rotations using legumes and green manure,[11] promoting agroforestry and farm forestry for better soil fertility, and increased land productivity to answer multiple needs at the farm level and beyond,[10] creating programmes to deal with the issues of tillage and depth of root bed to create sufficient storage

Table 1 Occurrence, characteristics, use and management of major soil types of the tropical region (Source: Palm *et al.*).[136]

Constraints	*Occurrence*	*Characteristics*	*Use and management*	*Soil types**
Soils with salinity problem.	Mainly in arid and semi-arid areas, on flat to very gently undulating topography and in poorly managed irrigation areas; surface and ground water may be the main source of soluble salts when used for irrigation; nature of the parent material can also be the source of soluble salts.	A saline soil profile has a pH below 8.5; white salt crystals are common on the sides of the soil profile or salt crust can be seen on the surface.	Can be reclaimed by leaching excessive soluble salts. Sometimes may be advisable to leave the soils under natural vegetation as it can be very expensive to reclaim them especially where availability of good quality water for irrigation is a limitation. Widely used for growing horticultural crops in small-scale irrigation schemes. However, salt accumulation in most of these schemes has led to their abandonment. This is because such soils require major land improvement and thus become extremely difficult to manage under low–input conditions which characterise African small-scale farmers.	Solonchaks (saline soils) or other soils classified as having a saline phase.

Table 1 (*Continued*)

Constraints	*Occurrence*	*Characteristics*	*Use and management*	*Soil types**
Soils with sodicity problem (note that saline – alkaline soils affect about 24% of the continent).	Mainly in arid and semi-arid areas, on flat to very gently undulating topography, brought about by prolonged or repeated saturation with water in which sodium predominates over the other cations.	The pH ranges from 8.5 to 10.0; soils have a light, coarse-textured topsoil over a compact, heavy-textured subsoil. Easily identified as they are characterised by columnar structure in the subsoil. Sodium dominates in the exchangeable complex, with exchangeable sodium percent (ESP) of more than 15%.	Highly advised to leave the soils under natural vegetation. Gypsum and heavy application of farmyard manure (FYM) are used to reclaim these soils. Soils with high sodium content are very prone to piping and gully erosion due to the poor structure and dispersion of clay when wet.	Solonetz (soils with a high content of exchangeable sodium) or other soils classified as having a sodic phase.
Soils with acidity problem.	Commonly referred to as leached tropical soils. Mainly occur in the highlands and mountainous areas as well as in the uplands with rolling to undulating topography with high precipitation, typical in coffee, tea and pyrethrum growing areas; these soils are exposed to excessive leaching.	Mostly characterised by light brown, red or reddish brown colours. Most of the soils under this category are well drained, deep with high clay content in the subsoils (except those classified as Ferralsols). Have low pH levels, usually <5.5; problem of phosphorus fixation due to high levels of aluminium; problems with soil micronutrients, especially molybdenum (Mo) are also common.	Apparently these are some of the most productive soils of the tropics if well managed. Internal drainage, workability and water-holding capacity are good. Application of lime, inorganic and organic fertilisers is necessary in order to maximise agricultural production, application of acidifying fertilisers to be avoided, with the exception of crops that thrive well under low pH, such as tea. The soils do respond well to liming, fertiliser (especially N and P) and FYM application. Erosion may be a serious problem as they occur in steep slopes, indicating that soil conservation measures must be in place. Mostly the soils are under numerous subsistence crops, tea, coffee, pyrethrum, temperate crops and dairy farming.	Andosols (young volcanic soils), Nitisols (deep, red, well-drained tropical soils with a subsoil horizon showing shiny ped surfaces), Acrisols (strongly weathered acid soils with low base saturation), Lixisols (strongly weathered soils in which clay is washed down from the surface soil to the underlying subsoils), Alisols (strongly acid soils with subsurface accumulation of activity clays that have more than 50% Al^{3+} saturation), Luvisols (soils in which clay is washed down from the surface soil to an accumulation horizon at some depth), Ferralsols (red and yellow tropical soils with a high content of sesquioxides).

Table 1 (*Continued*)

Constraints	*Occurrence*	*Characteristics*	*Use and management*	*Soil types**
Soils with drainage problem	Occur in relatively flat areas such as alluvial plains, depressions, river valleys and valley bottomlands that are subject to waterlogging during wet seasons; they are found in both semi-arid and sub-humid regions; in particular, soils classified as Fluvisols and Gleysols are developed on a wide range of unconsolidated materials, mainly sediments.	These are generally poorly drained soils with very slow vertical drainage; the colour ranges from grey to black; workability is extremely poor; root development is hampered by oxygen deficiency during the wet periods; population of soil microbes is low due to prolonged waterlogging; other soils in this category, such as Vertisols (black cotton soils), have the capacity to expand (swell when wet) and shrink (when dry) thus affecting crop growth.	Main obstacle to the utilisation of these soils is the necessity to install a drainage system. Broad – bed and furrow systems have proved successful in the management of these soils, especially those classified as Planosols; furrows should have very slight slopes to avoid gullying; large areas under Vertisols are used for growing paddy rice, sugar cane, cotton, teff and subsistence crops while grazing is common during the dry seasons; other soils under this category, such as Gleysols occurring along the river valleys are utilised during the dry seasons for growing vegetables and horticultural crops which have a higher market value at such times. Those classified as Histosols occur mainly in wetlands and are therefore best kept under a permanent grass cover or under swamps or forest.	Planosols (vlei soils with light textured topsoil abruptly over dense subsoil), Vertisols (black cotton soils), Gleysols (soils with clear signs of excess wetness), Fluvisols (soils developed in alluvial deposits) and Histosols (peat and muck soils).

Table 1 (*Continued*)

Constraints	*Occurrence*	*Characteristics*	*Use and management*	*Soil types**
Soils with soil fertility problem.	Variable – virtually all soils in sub-Saharan Africa have fertility problem.	Low fertility and insufficient organic matter are the main soil characteristics that adversely affect crop production in the region; most of the soils are highly weathered, having undergone both ferrugination and ferralitisation processes, have low pH and high levels of aluminium toxicity which interfere with nutrient uptake; nitrogen (N) and phosphorus (P) are the most limiting nutrients and so are the low levels of organic matter; low soil fertility results from nutrient depletion – erosion, crop harvest leading to negative nutrient balance, leaching and inadequate crop fertilisation.	Used for extensive grazing and for growing a wide range of subsistence and cash crops, both annual and perennial. Requires heavy application of both organic and inorganic fertilisers including liming, depending on levels of soil pH. Appropriate agronomic practices, soil and water conservation measures are required for soil moisture conservation as well as for minimising water erosion (see Figure 1).	A wide range of soils in the region fall under this category. However, those classified as Ferralsols are naturally chemically poor as most of the primary minerals have weathered, resulting in soils having low cation exchange capacity (CEC) of <16 meq. Thus maintaining the SOM content by manuring, mulching or adequate fallow period and prevention of erosion are important management requirements.

Table 1 (*Continued*)

Constraints	*Occurrence*	*Characteristics*	*Use and management*	*Soil types**
Soils with root restriction layers (note that the term 'effective rooting depth' is used to include chemical barriers which reduce the volume of the soil for root exploitation.	Variable – found in all climatic zones in sub-Saharan Africa.	May be as a result of shallow ground water table (as is the case with soils classified as Gleysols or Histosols) or physical due to the presence of rock or other cemented materials such as plinthite and rock.	Erosion is the greatest threat to these soils. Soil moisture storage capacity may be limiting for most of the crops for soils with rooting depth limitations due to presence of rock at shallow depth. Where rooting depth is as a result of shallow ground water table, the soils can be made productive by putting in drainage measures (see above for soils with drainage problem). Thus, depending on the type of crops to be grown, such soils can be quite productive (see Figure 2).	Gleysols and Fluvisols (if having a shallow ground water table); Leptosols (*shallow soils*), Regosols (*soils in the weathered shell of the earth*), Plinthosols (*soils with plinthite*) and other soils classified as having a lithic or plinthic phase.

*Soils classified according to FAO-UNESCO Soil Classification System. (Source: Palm *et al.*).[136]

Figure 1 Intercropping (left) in an orthic Ferralsol (right). Sometimes poor agronomic practices have led to poor crop growth. Certainly the maize crop in this farm lacks nitrogen. Competition for resources such as nutrients is common under this kind of cropping system with no fertiliser inputs. (Source: Gachene[9] (Gachene and Kimaru)).

capacity for plant nutrients and water, especially for soils with a compacted sub-soil. Further issues of the required energy and the development of new or improved tillage systems and equipment need to be dealt with as crucial

Figure 2 With proper soil and water management practices, a shallow profile like the one on the left can be made productive. This soil, when well mulched, can support a good crop of tomatoes and palm trees as shown in the right photo. The use for which the soil is been assessed is critical in land evaluation. (Source: Gachene[9] (Gachene and Kimaru)).

Table 2 Physical and chemical topsoil properties at Fertiliser Use Recommendation Project trial sites in the Kenyan highlands.

Location	*Altitude (m)*	*Soil type*	*Texture class*	*Bulk density* ($g\ cm^{-3}$)	*pH* H_2O	*Org C* ($g\ kg^{-1}$)	*Total N* ($g\ kg^{-1}$)
Otambo	1790	Mollic Nitisol	C	0.89	6.4	29	3.4
Kiamokama	2020	Humic Nitisol	C	0.91	6.2	25	3.3
Kisii NARLS	1730	Mollic Nitisol	C	1.19	6.1	23	2.8
Rodi Kopany	1330	Pellic Vertisol	C	–	6.6	16	1.9
Rongo	1440	Humic Acrisol	LS	1.42	6.3	7	0.9
Oyugi Ober	1450	Chromo – luvic Phaeozem	C	1.25	6.4	18	2.1
Ukwala	1200	Orthic Acrisol	SCL	1.63	5.7	5	0.8
Siaya Obambo	1200	Chromic Luvisol	C	1.32	6.0	14	1.7
Busia Buburi	1220	Ferralo- chromic Acrisol	C	1.26	5.0	11	1.6
Kamakoiwa	1710	Rhodic Ferralsol	SC/C	1.19	5.5	13	1.6
Tongaren	1725	Ferralo- chromic Acrisol	C	1.13	6.4	14	1.0
Mumias	1270	Ferralo- chromic Acrisol	SL	1.61	5.1	5	0.6
Kakamega NARS	1520	Dystromollic Nitisol	C	1.20	5.6	24	2.3
Vihiga Malagoli	1620	Nitohumic Ferralsol	C	1.28	5.3	27	2.8
Baraton	2000	Humic Nitisol	CL	0.95	5.6	36	3.6
Chepkumia	1750	Humic Acrisol	CL	1.15	5.5	28	4.5
Sosiot	1890	Dystromollic Nitisol	C	0.89	5.6	40	5.4
Chebunyo	1840	Vertieutric Planosol	CL	1.12	6.9	43	5.4
Kitale NARS	1860	Humic Ferralsol	SC	1.42	6.2	16	2.0
Eldoret	2140	Ferraric Cambisols	C	–	5.0	13	1.1
Turbo	1850	Ferralo- chromic Acrisol	SC	1.34	5.0	15	1.1
Kapenguria	2140	Humic Cambisol	SC	0.87	7.5	32	3.1
Bugar	2320	Humic Nitisol	C	1.06	5.2	19	1.9

Table 2 *(Continued)*

Location	*Altitude (m)*	*Soil type*	*Texture class*	*Bulk density ($g\ cm^{-3}$)*	*pH H_2O*	*Org C ($g\ kg^{-1}$)*	*Total N ($g\ kg^{-1}$)*
Kasoiyo	1990	Dystric Nitisol	C	1.10	6.1	20	1.7
Eldama Ravine	2100	Nitochromic Luvisol	C	0.90	5.6	25	1.9
Ol Ngarua	1970	Nitoferric Luvisol	C	1.18	6.0	23	2.6
Upepo farm	2180	Chromic Luvisol	C	1.11	5.7	22	2.4
Rotian	2180	Luvic Phaeozem	CL	1.12	5.7	25	2.3
Ol Joro Orok Charagita	2780	Nitochromic Luvisol	CL	0.98	6.2	34	4.3
Ol Joro Orok ARS	2360	Andoluvic Phaeozem	C	1.14	6.3	27	3.0
Tulaga	2530	Eutric Planosol	CL	1.12	6.0	21	1.5
Njabini	2530	Andoluvic Phaeozem	CL	1.01	6.3	32	4.6
Githunguri	1930	Humic Nitisol	C	0.85	5.9	19	3.3
Kandara	1640	Humic Nitisol	C	1.08	5.3	23	1.9
Makuyu	1430	Dystric Nitisol	C	1.07	5.3	15	1.2
Muirungi	2080	Andohumic Nitisol	CL	0.85	5.7	42	4.7
Chehe	1920	Andohumic Nitisol	CL	0.71	4.9	27	3.3
Kerugoya	1480	Humic Nitisol	C	1.20	5.6	29	3.4
Kavutiri	1700	Andohumic Nitisol	C	0.77	4.7	36	3.9
Gachoka	1200	Rhodic Ferralsol	C	1.07	5.7	17	1.4
Embu ARS	1510	Humic Nitisol	C	0.99	5.8	21	2.4
Kaguru	1460	Humic Nitisol	C	0.91	5.0	11	1.6
Kilome Upepo	1680	Rhodic Ferralsol	SCL	1.13	6.9	14	1.4
Makutano	1310	Orthic Acrisol	LS	1.51	6.1	6	0.6
Weruga	1690	Chromic Acrisol	SC	1.22	5.6	14	1.7

Table 3 Some chemical characteristics of selected sites in West Africa. (Source: Sy).[81]

Site	*Soil classification*	*Organic matter (%)*	*Soil pH (H_2O)*	*CEC (cmol kg^{-1})*	*Available P (bray 1) (mg kg^{-1})*
Fendal (Liberia)	Plinthic Acrisol	1.5	5.0	1.0	6
Owem (Nigeria)	Acrisol	2.2	4.8	5.2	6
Kwadaso (Ghana)	Acrisol	1.3	4.9	3.5	2.2
Samaru (Nigeria)	Lixisol	1.0	5.8	4.3	3.5
Davie (Togo)	Nitisol	0.8	6.0	2.8	1.4
Kaboli (Togo)	Lixisol	1.1	5.9	2.4	1.2
Farakoba (Burkina Faso)	Lixisol	1.0	5.4	0.8	2.7
Agonkamey (Benin)	Alfisol	0.6	6.6	2.3	2.0
Yundum (Gambia)	Lixic Ferralsol	1.1	5.5	8.1	15.2
Saria (Burkina Faso)	Arenosol	0.6	5.3	1.8	2.5
Gaya (Niger)	Arenosol	0.7	6.3	1.7	2.3
Sadore (Niger)	Aridic Arenosol	0.3	5.0	1.0	2.8
Sotuba (Mali)	Lixisol	0.5	5.4	2.3	1.7

elements in the process. Such improved methods of tillage should lessen the problem of hardpans and plough soles. This will greatly enhance soil water uptake for plant growth,[13] developing efficient systems of irrigation that increase production without degrading the soil,[14] and adopting soil conservation measures that are simple, effective and affordable.[15] Thus, understanding the soil is the key to its improvement as there are many physical, chemical and biological properties of the various soil types that affect plant growth.

4 Soil Biology: Role of Soil Biodiversity and Functions (Ecosystem Services)

Soil biota are an essential component of soil health and constitute a major fraction of global terrestrial biodiversity.[16] Within the context of Integrated Soil Fertility Management (ISFM), soil biota are responsible for the key ecosystem functions of decomposition and nutrient cycling, soil organic matter synthesis and mineralisation, soil structural modification and aggregate stabilisation, nitrogen fixation, nutrient acquisition, regulation of atmospheric composition, the production of plant growth substances and the biological control of soil-borne pests and diseases.[17] Understanding biological processes is not as well advanced as those that are related to soil physical and chemical properties, creating opportunities for breakthroughs in biotic function to better service agriculture. These services accrue through two basic approaches: indirectly, as a result of promoting beneficial soil biological processes and ecosystem services through land management, or directly through the introduction of beneficial organisms to the soil.[18] Soil macrofauna, especially earthworms and termites, are important components of the soil ecosystem and as ecosystem engineers they influence formation and maintenance of the soil structure and regulate soil processes. Earthworms and termites have different feeding strategies which, in turn, affect their impact on soil. Because of their sensitivity to disturbance and their importance in redistributing and transforming organic inputs, earthworms and termites represent an important indicator of soil quality.

Soil invertebrates are important determinants of biological, chemical and physical characteristics. They enhance biodegradation and humification of organic residues in several ways: (1) by breaking down organic residues and increasing surface area for microbial activity; (2) by producing enzymes which break down complex bio-molecules into simple compounds to form humus; and (3) by improving the soil environment for microbial growth and soil-plant interactions.[19–21]

The diversity and abundance of the structures produced by soil ecosystem engineers, *e.g.* earthworms and termites, impact on the physical properties of soils, *i.e.* overall aggregation, porosity, water infiltration and retention and resistance to erosion.[22] Earthworms play an important role in the formation of soil organic matter (SOM) enriched macroaggregates,[23–26] which can physically protect occluded organic matter against microbial decay and, upon disintegration, release occluded carbon and nutrients.[23,27] Apart from promoting soil physical and chemical properties, earthworms also promote nodulation,[28] dispersal of mycorrhizal fungi,[29] and even disease suppression and dispersal.[30] Termites mediate the synthesis and breakdown of soil organic matter and influence water infiltration and availability to plants by modifying soil structure.[31–35] They influence soil physical properties through the construction of mounds, nests, galleries and surface sheeting[31,34,36] and also by transporting materials, thereby producing passages which improve drainage and aeration.[37–39] Mound-building termites form stable microaggregates that physically protect occluded organic matter against rapid decomposition and reduce soil erosion and crust formation.[40,41]

The importance of termites in the decomposition of plant matter in natural ecosystems is well documented;[42–45] it has been established that in the tropical rainforests of Nigeria termites play a significant role in both decomposition and litter removal. Mando and Brussard[42] found that termites alone could account for up to 80% of litter disappearance in one year. Termites play a significant role in soil nutrient availability and cycling through interactions with other soil organisms, *e.g.* bacteria and fungi, to most of which they provide food.[40] Soil from termite mounds is sometimes used as fertiliser in tropical cropping systems because of a high accumulation of nutrients.[46,47] Despite the potentially beneficial role of termites, termite pest problems have been identified as a major constraint to increasing yields of crops in sub-Saharan Africa.[48,49] The challenge therefore remains to better understand the interactions between agricultural management practices and soil fauna (*e.g.* termites) in order to find ways to enhance soil fertility and crop yields.

Soil microorganisms are a source of important medicines, including most of the early antibiotics such as penicillin. But despite their functional importance, the soil biota remain a "black box" to scientific understanding as well as to the common gaze due to a number of challenges which include lack of appropriate methods to study this myriad of organisms and their complex ecosystem. The role they play in determining some crucial ecological functions has resulted in a shift in the way scientists view them and there is a major attempt to amass knowledge so as to exploit them for development of sustainable utilisation and management of soil resource. It is against this background that the Global Environment Facility-United Nations Environment Programme (GEF-UNEP)-funded global project on the conservation and management of below-ground biodiversity (CSM-BGBD) was conceived.

5 Case Studies: Effect of Management and/or Land Use Intensification

5.1 Soil Carbon as Fuel for Soil Organisms

Maintenance of soil organic matter (SOM) through integrated soil fertility management is important for soil quality and agricultural productivity, and for the persistence of soil faunal diversity, abundance and biomass. In turn, soil macrofauna affect SOM dynamics through organic matter incorporation, decomposition and the formation of stable aggregates that protect organic matter against rapid decomposition.

Integrated soil fertility management (ISFM), widely advocated in sub-Saharan Africa, recognises the benefits of combining organic and inorganic fertilisers for sustainable nutrient management.[51,52,56] The beneficial effect of soil organic matter (SOM) on soil productivity through supplying plant nutrients, enhancing cation exchange capacity, improving soil aggregation and soil and water retention, is well established.[53–55] In addition, SOM supports various soil biological processes by being a substrate (source of carbon) for decomposer organisms and ecosystem engineers, such as earthworms and termites that play an important role in soil structure formation, organic matter decomposition and nutrient mineralisation.[53,54] Ayuke *et al.*[56] showed that arable cropping has significant negative effects on earthworm, but little effect on termite diversity as compared to long-term fallow. Under continuous crop production, higher earthworm and termite diversity was observed under agricultural management that had resulted in high-C *versus* low-C soils.

To reiterate the benefits of ISFM as promoter of soil biodiversity, Ayuke *et al.*[57] demonstrated that long-term application of manure in combination with fertiliser result in higher earthworm taxonomic richness and biomass (see Table 4), which leads to improved soil aggregation and enhanced C and N stabilisation within this more stable soil structure.[57] It is possible that the long-

Table 4 Earthworm and termite taxonomic richness, mean count (number) and biomass (in parentheses) per monoliths (0.0625 m^{-2}) of the 0–30 cm soil layer of a Humic Nitisol under different residue, manure and mineral fertiliser management at the Kabete field trial, Kenya.

Treatment		*C-F*	*C+F*	*R-F*	*R+F*	*FYM-F*	*FYM+F*	*NF*
Organic fertiliser (OF)		None	None	Residue	Residue	Manure	Manure	na
Mineral fertiliser (MF)		?	+	?	+	?	+	na
Taxonomic group	**Functional group**[a]							
Earthworm Ocnerodrilidae								
Nematogenia lacuum	Endogeic	17 (0.3)	13 (0.2)	5 (0.1)	5 (0.1)	7 (0.1)	23 (0.5)	14 (0.2)
Gordiodrilus wemanus	Endogeic	?	?	?	?	?	?	1 (0.01)
Acanthodrilidae								
Dichogaster (Dt.) affinis	Epigeic	?	?	?	?	?	1 (0.05)	1 (0.02)
Dichogaster (Dt.) bolaui	Epigeic	?	?	?	?	?	1 (0.04)	3 (0.1)
Eudrilidae								
Polytoreutus annulatus	Epigeic	?	?	?	?	?	?	4 (0.2)
Stuhlmannia spec nov	Epigeic	?	?	?	?	?	?	1 (0.01)
Species richness (S)		1	1	1	1	1	3	6
Termites Termitidae-Macrotermitinae								
Microtermes spp.	G II (FWLG)	1 (0.1)	10 (0.01)	3 (0.0)	?	?	24 (0.1)	3 (0.0)
Odontotermes spp.	G II (FWLG)	4 (0.0)	6 (0.01)	18 (0.1)	?	9 (0.0)	25 (0.4)	60 (0.0)
Pseudacanthotermes spp.	G II (FLSD)	11 (0.0)	?	?	7 (0.0)	?	?	?
Genus richness (S)		3	2	2	1	1	2	2

C−F = control minus fertiliser, C+F = control plus fertiliser, R−F = residue minus fertiliser, R+F = residue plus fertiliser, FYM−F = farm yard manure minus fertiliser, FYM+F = farm yard manure plus fertiliser, NF = natural fallow, n.a. = not applicable (Source: Ayuke *et al.*).[56]

term application of combined farmyard manure and fertiliser (FYM +F) resulted in increased soil C concentration, providing food sources for earthworms and mulching effect on their habitat and also stimulating plant growth and litter return,[58] resulting in higher earthworm biomass.

5.2 Soil Macrofauna in Tropical Agroecosystems

In large parts of sub-Saharan Africa (SSA), pests, weeds, diseases and soil fertility decline are major biophysical causes of low *per capita* food production.[59] Degradation processes, such as loss of soil carbon and nutrient depletion in general, can occur quickly and are difficult to reverse.[60] Moreover, loss in yield cannot be corrected by the use of fertilisers in economies where cash flow is minimal. Under such circumstances, Integrated Soil Fertility Management (ISFM), *i.e.* integration of fertilisers with organic resources, has been regarded as a feasible alternative in low-input systems, compensating for the high costs of fertilisers.[52] Manipulation of the soil environment *via* tillage, application of organic residues and manipulating soil fauna are among the factors affecting SOM dynamics under cropping systems.[61,62] In low-input agricultural systems, soil fauna have been found to play a crucial role in soil organic matter dynamics, in soil physical properties improvement, and in nutrient release for crop production.[63] However, soil macrofauna composition, abundance and activity, and hence their impacts on soil processes, vary depending on residue inputs and soil management practices.[25,64,65] Climate, soil texture and management have been indicated to influence the activity of soil macrofauna (*e.g.* earthworms and termites) that produce biogenic structures.[66] It can therefore be postulated that differential land-management effects on soil fauna functional groups can translate into differential effects on the structures they produce, thus affecting soil organic matter, soil aggregation, porosity and water and nutrient availability to plants.

Figure 3 shows a hierarchical model of inter-correlated factors that determine soil biodiversity and processes. Management practices (*e.g.* crop rotation, tillage, organic resource use and application of agrochemicals such as pesticides, herbicides and inorganic fertilisers) can cause positive or negative changes in species composition, community structure and population sizes. Some of the negative effects of management practices may be long-lasting and result in a decline in the abundance and/or biomass of soil macrofauna populations, or eliminate or reduce key species, *i.e.* species that play a disproportionate role in ecosystem processes.[67,68] The use of organic inputs and crop diversification through rotation favours macrofauna diversity due to improvement in the abiotic conditions and increased substrate supply.[58,69–71] Agroforestry technologies, such as alley cropping, natural fallows (bush fallows), planted fallows and biomass transfer systems can restore activities of organisms such as earthworms, termites, ants and other microarthropods.[72–74] Ayuke *et al.*[69,72] found that organic residues from *Senna spectabilis* and *Tithonia diversifolia* increased the population of earthworms by 400% and

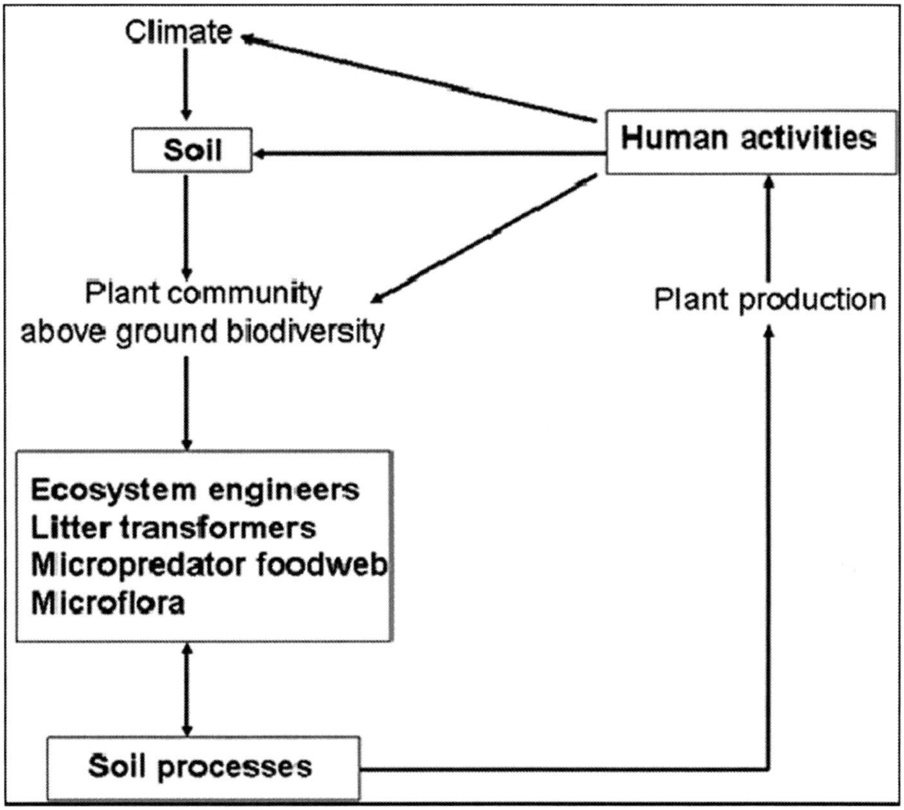

Figure 3 A hierarchical model of factors that determine soil biodiversity and soil processes.[161]

240% over a no-input control, respectively, while termites increased by 150% and 120% when the two different organic residues were added to the soil (see Table 5). Tian *et al.*[71] similarly found higher earthworm and microarthropod populations under planted fallows than under continuous cropping systems

Table 5 The abundance of soil macrofauna under different treatments in soil at Maseno, Western Kenya.

	Earthworms	*Termites*	*Other macrofauna*
Treatment		*Number (m^{-2})*	
Control	99 (9.5) c	229 (14.0) b	43 (6.3) d
Fertiliser	132 (11.0) c	348 (16.0) b	90 (9.4) c
S. spectabilis	572 (22.5) a	737 (25.4) a	391 (15.5) b
T. diversifolia	339 (18.5) b	652 (21.4) ab	309 (14.0) b
SED	(1.0)	(6.3)	(1.1)

Means followed by the same letter in a column are not significantly different at $p < 0.05$. Values in parenthesis are square root transformed. SED = Standard error of difference of means. (Source: Ayuke).[72]

and attributed this to higher litter fall, lower temperature and higher soil moisture.

5.3 Mesofauna

Mesofauna includes organisms less than 4 mm long (or 2 mm wide). They mostly live in the litter or soil cracks and pores. Examples are the micro-arthropods, mites, springtails, enchytraeidae, *etc.*

The structure, organisation and behaviour of individuals within soil fauna communities dynamically respond to seasonal and diurnal changes in environmental conditions. In addition, the distribution of individuals in space is heterogenous within a given habitat. Variation in environmental conditions, biotic interactions and colonisation history result in uneven distribution of soil fauna in space. As such, management practices that alter the environmental conditions are likely to have greater impact on the diversity of mesofauna groups as well.[74]

In a maize-based system of western Kenya, faunal composition and abundance within the agroecosystem were dominated by macrofauna groups (90.2%), while mesofauna groups constituted only 9.8%.[72] Maribe *et al.*[75] monitored the abundance and diversity of mesofauna groups such as mites along a gradient of land-use types in Taita Taveta, Kenya. They found that land-use types significantly influenced the abundance and diversity such that intensification lowered the diversity and abundance, resulting in a less complex mite community structure (see Table 6). Higher abundance, richness and diversity were observed in less disturbed forest ecosystems unlike the agroecosystems, which are often disturbed during cultivation.[76]

In another study, Birgit *et al.*[76] found that application of organic amendments such as cow manure encouraged proliferation of collembolan

Table 6 Mean abundance, richness and diversity of soil mites at Taita Taveta during long rains in April 2008.

Land use types	*Mean abundance*	*Mean richness*	*Shannon-Wiener index*
Maize-based	72.3 ±24.7d	6.5 ±1.9c	1.3 ±0.3bc
Coffee	120.5 ±25.7d	10.8 ±1.1bc	1.8 ±0.1ab
Horticulture	132.3 ±22.7d	6.0 ±1.1c	1.1 ±0.3c
Napier	147.7 ±70.1cd	8.7 ±2.3bc	1.1 ±0.3c
Natural forest	244.0 ±63.3bcd	12.3 ±0.9ab	2.1 ±0.1a
Fallow	413.8 ±79.4abc	12.0 ±2.9ab	1.1 ±0.2c
Pine forest	436.2 ±181.7a	15.8 ±1.6a	2.0 ±0.2a
Cypress forest	607.0 ±118.8a	16.8 ±1.1a	2.2 ±0.2a
F test	$F_{7,23}=4.51$; $P = 0.003$	$F_{7,23}=5.50$; $P < 0.001$	$F_{7,23}=5.57$; $P < 0.001$

Means followed by the same letters within a column are not significantly different at $p < 0.05$ (Fisher test). (Source: Maribe *et al.*).[75]

population, whereas inorganic fertilisers negatively impacted on these organisms.[77]

5.4 Beneficial Microorganisms: Soil Fertility Promoters, Plant Growth Regulators and Biocontrols

Soil ecosystems are among the most complex of all terrestrial communities, and the role of the soil biota in maintaining plant health is progressively being understood. The composition of the soil biota is strongly influenced not only by the nature of the underlying organic matter and mineral components, but also by environmental variables such as temperature, pH and moisture. Natural soils have been shown to harbour large populations of microorganisms which exist in a state of dynamic equilibrium and controlled changing balances. These microorganisms primarily compete with each other for nutrition and space. A majority of the microbes are classified as fungi and bacteria which play beneficial and often vital roles in natural environments and agriculture. Numerous benefits are accrued from these microbes including (1) direct symbiotic association with roots (mycorrhizae, legume nodulating bacteria); (2) nutrient cycling which involves breakdown and release of minerals from organic matter present in the soil, resulting in increases in essential element availability to higher plants; and (3) biocontrol agents, through predation of disease-causing microorganisms and/or suppressing growth, or reproduction activity of harmful disease-causing microorganisms through other interactions such as chemical inhibition. Details of selected microbes with economic potential which have been well investigated in African soils are discussed in the sections below.

5.4.1 Legume Nodulating Rhizobia (LNB)

Biological nitrogen fixation is the ability of living organisms to convert inert dinitrogen gas in the atmosphere (N_2) into nitrogen-containing organic compounds through asymbiotic, associative or symbiotic processes. Microbially mediated nitrogen fixation accounts for 175 million tonnes per year in terrestrial and aquatic environments.[78] This provides two thirds of the nitrogen required in the biosphere, most of which comes from the contribution of the association between modulating rhizobia bacteria with compatible host legumes.

The organisms that possess the nitrogenase enzyme have attracted considerable interest. These prokaryotes in the Eubacteria and Archaebacteria kingdoms which can fix nitrogen are metabolically diverse and the different bacterial N-fixing systems have been reviewed.[78] For almost 100 years the term *Rhizobium* was used to represent those organisms capable of forming nodules with specific homologous host legumes. Recently, phylogenetic analysis which uses 16S rRNA has become the standard for classification of bacteria. This new classification, which is dependent on the phenotypic traits, has confirmed a number of taxonomic divisions which include Azorhizobium, Bradyrhizobium,

Mesorhizobium and Rhizobium.[79,80] The technique has been used in numerous studies of African soils which have revealed rhizobia diversity of the LNB in African soils. For instance, identification of the genus *Methylobacterium* in Senegal by Sy[81] and Samba *et al.* [82] reported a total of 117 strains of both slow- and fast-growing rhizobia from roots of Crotolaria species in Senegal. Similarly, Odee *et al.*[83] identified five bacteria genera, namely *Agrobacterium, Bradyrhizobium, Mezorhizibium, Rhizobium and Sinorhizobium,* for root nodules of legumes growing in diverse soils in Kenya, while Anyango *et al.*[84] found that beans grown in acid soils in Kenya were nodulated by different rhizobia species. In a recent study which assessed the abundance of LNB in soils of the Embu and Taita Districts in Kenya, Mwenda *et al.*[85] and Mwangi *et al.*[86] obtained similar rhizobia diversity to Odee *et al.*[83] and their diversities were positively influenced in cropping systems.

Legume inoculation is a process through which leguminous crops are provided with the effective bacterial strain of the genus *Rhizobium* which results in an effective symbiotic relationship that brings about fixation of atmospheric nitrogen into organic nitrogenous compounds in the plant. However, response to rhizobia inoculation is influenced by a number of factors which include soil nitrogen, rhizobia strain and populations of indigenous populations, crop and environmental conditions.[87] Despite these challenges, inoculant production has going on for several decades by both private and public institutions in Africa to harness benefits of the Legume-Rhizobium technology and about 100 000 tonnes of rhizobia inoculants are produced in Kenya, South Africa, Zambia and Zimbawe for inoculating food legumes such as soya bean, beans and also for fodder crops.[87]

5.4.2 Arbuscular Mycorrhizal Fungi (AMF)

Arbuscular mycorrhizal fungi (AMF) are common root-colonising fungi forming symbioses with most plants. The AMF are globally widespread and are associated with most plant species.[88,89] These fungi have been reported from diverse natural ecosystems including deserts, sand dunes, tropical forests, salt marshes, and in managed systems such as pastures, orchards and field crops.[90] In agricultural systems, edaphic factors, land use, cropping systems and management practices interact to influence AMF species composition and spore population. Consequently, changes in agricultural practices will inevitably lead to a change in the overall abundance of propagules of each fungus within a population.[90] Studies carried out on the distribution of AMF in legume-based systems in Nigeria showed prolific arbuscular mycorrhizal colonisation in the roots.[91] Shepherd *et al.*,[92] on the other hand, found forest and grassland soils to have narrower species distribution than most farm soils, indicating some degree of ecosystem adaptation. In a survey carried out in the Mount Kenya region, across different land-use types (LUTs), a total of 16 AMF species were isolated.[93] The spore community was dominated by Acaulosporaceae and Glomaceae. Land-use type had no significant effect on

AMF spore abundance or root colonisation. Trends, however, showed soils under napier (*Pennisetum purpureum* Schumach) and tea (*Camellia sinensis* L.) had the highest AMF spore abundance while natural forest and planted forest had the least spore abundance (see Figure 4). The reverse was observed for root colonisation where the highest colonisation was in soils under natural and planted forest, except tea which maintained both high spore abundance and slightly high colonisation.

Infection of crop roots with AM fungi can improve the uptake of nutrients, particularly phosphorus, and increase crop production.[94] These endomycorrhizal fungi are obligate symbiotic fungi, the hyphae of which develop mycelia, arbuscules, and in most fungal genera vesicles in roots. Soil hyphal networks produced by these symbiotic fungi provide a greater absorptive surface area than plant root hairs. As such, mycorrhizal symbiosis assists crops in recovering scarce reserves of soil phosphorus. In addition, mycorrhizal-infected plants have been shown to have greater tolerance to toxic metals, root pathogens, drought, high temperatures, saline soils, adverse soil pH and transplant shock than non-mycorrhizal plants.[95] Mycorrhizal association has been recognised for cassava production, given that it is usually grown in infertile soils, without fertiliser application.[96] Inoculation of orange-fleshed sweet potato varieties with mycorrhizal fungi and phosphate-solubilising bacteria (PSB) in the low-phosphorus soils increased phosphorus concentration in the soil and root yield. Arbuscular mycorrhizal fungi therefore constitute one of the strategic interventions for ISFM. Two basic strategies to manage mycorrhizal fungi are available through optimising crop and

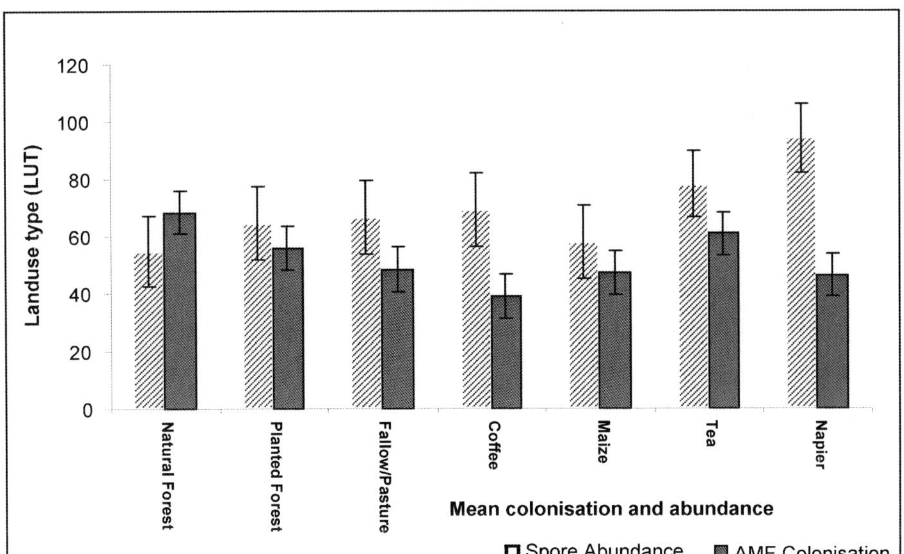

Figure 4 Impact of land use type (LUT) in order of less-to-high intensity on spore abundance and colonisation. (Source: Jefwa *et al.*).[93]

management practices that affect the abundance of indigenous mycorrhizae, or through the use of mycorrhizal inoculants.[90] While it has become widely accepted that mycorrhizal symbiosis, in combination with legumes, can be harnessed to improve crop productivity, maximise root functions, and also reduce fertiliser use, there is still need to establish the distribution and functions of AMF species in different habitats and different land-use systems in order to facilitate inoculation programs. With improved methods and technologies in utilisation, approaches to studying AMF should be streamlined in order to derive maximum benefits from the association.

Although ectomycorrhizae have not been given much attention in agroecosystems, they are equally crucial in afforestation programmes.[97] Through hyphae, nutrients and water can be absorbed by trees. Ectomycorrhizae are mostly found in woody plants, ranging from shrubs to forest, and many belong to the families Pinaceae, Fagaceae, Butulaceae, Casuarinaceae and Myrtaceae. Most of the above host plants are specific, such that if mycorrhizae are absent growth is highly reduced.[97] Over 4000 species of Basidiomycotina and a few Ascomycotina form ectomycorrhizae. Many of these fungi produce mushrooms and puffballs on forest floor.

5.4.3 Plant Growth Promoting Rhizobacteria (PGPR)

Beneficial free-living soil bacteria are referred to as plant growth promoting rhizobacteria or PGPR and they stimulate plant growth either directly or indirectly through secretion of phytohormones that enhance plant growth or uptake of solubilised iron from the soil.[98] Solubilisation of nutrients such as phosphorous through production of organic acids releases insoluble phosphorus into more soluble forms.[99,100] Paterno[101] concluded from his study that *Azotobacter vinelandii* and *Bacillus cereus* produced high amounts of indole acetic acid (IAA). Karawal[102] reported that the 30 isolates of *Pseudomonas fluorescens* he tested were indole-positive, indicating production of IAA. However, the *Pseudomonas fluorescens* showed higher IAA production when tryptophan concentrations were increased. Gachie (unpublished,University of Nairobi, 2012), in her screening experiment of rhizobacteria (40 isolates of *Bacillus* spp, 36 isolates of *Azotobacter* spp and 53 isolates of *Pseudomonas* spp), all from soils collected from potato-producing districts in Kenya, identified rhizobacteria isolates with plant growth promoting, phosphorus solubilisation potential, while other isolates controlled the *Ralstonia solanacearum* potato pathogen which is widespread in Kenyan soils and is a major constraint to growth of the potato industry.

5.4.4 Trichoderma

Trichoderma species are cosmopolitan fungi found in decaying wood and vegetable matter. Their dominance in soil may be attributed to their diverse metabolic capability and aggressive competitive nature.[103] They colonise roots, attack, parasitise and gain nutrition from other fungi, thus enhancing root growth. They have developed rhizosphere competence through numerous mechanisms for attacking other fungi and for enhancing plant and root growth. These properties include mycoparasitism, antibiosis, competition for nutrients or space, tolerance to stress through enhanced root and plant development, solubilisation and sequestration of organic nutrients, induced resistance, and inactivation of enzymes.[104–106] A study conducted in two benchmark sites of Embu and Taita in Kenya yielded a total of 299 and 309 *Trichoderma* isolates, respectively,[107] and the most frequently isolated and abundant species from both sites was *T. harzianum*.

Trichoderma fungus has a high potential for the biological control of fungal root pathogens that can improve plant growth in infested soils.[105] Plants not infected with root pathogens often demonstrate a positive growth response after being treated with *Trichoderma* as well, suggesting production of a growth stimulant. A study by Okoth *et al.*[108] showed that *Trichoderma* inoculation significantly increased the rate of maize seed germination, further corroborating its potential as a growth stimulant. Recently, this fungus was commercialised as a soil inoculant and seed treatment of agricultural crops, with numerous commercial products being registered around the world.[105] *Trichoderma* species have been investigated for over 70 years.[104] They have been used as biological control agents (BCAs) and their isolates recently have become commercially available.[105] This development is largely the result of a change in public attitude towards the use of chemical pesticides and fungicides such as methyl bromide.[109,110] In this respect *Trichoderma* species have been studied as BCAs against soil-borne plant pathogenic fungi.[111,112] Replacement or reduction of chemical application can be achieved through use of biologically based fungicides, a concept included in the broad definition of biological control proposed by Cook and Baker.[113] Species in the genus *Trichoderma* are important as a commercial source of several enzymes and as biofungicides/growth promoters. The most common biological control agents of the genus are strains of *T. harzianum*, *T. viride* and *T. virens*. In a study in which sixteen selected isolates of *T. harzianum* from different land use types in Embu, Kenya, were tested for antagonism against five soil-borne phytopathogenic fungi (*Rhizoctonia solani*, *Pythium sp*, *Fusarium graminearum*, *F. oxysporum f. sp phaseoli* and *F. oxysporum f. sp lycopersici*) results showed that all *T. harzianum* isolates had considerable antagonistic effect on mycelial growth of the pathogens in dual cultures compared to the controls.[114] Since all *T. harzianum* isolates evaluated were effective in controlling colony growth of the soil-borne pathogens, both in dual cultures and in culture filtrates, they offer good prospect as broad spectrum biological control agents in the greenhouse and under field conditions.

5.4.5 *Bacillus subtilis*

Several strategies, including chemical nematicides, organic soil amendments, crop rotation, cover crops, resistant cultivars and biological control, have been developed for the management of plant parasitic nematodes.[115] Evidence has been provided that integrating biological control, using microbial antagonists with other possible methods, is amongst the most pragmatic strategies for managing nematodes. Biological control agents that have been assessed include egg-parasitic fungi, nematode-trapping fungi, bacteria and polyphagous predatory nematodes. Plant-growth-promoting rhizobacteria, especially the genera *Pseudomonas* and *Bacillus*, have demonstrated potential for disease suppression without negative effects on the user, consumer, or the environment. Some strains of *Bacillus subtilis* have exhibited potential as biocontrol agents in the management of root-knot nematodes.[116] In a study conducted at Kakamega County, Western Kenya, it was observed that *Bacillus subtilis* strains K158, 194 and 263 reduced the population of *Meloidogyne sp* in the following order: K158 >K263 >K194 (see Figure 5). Dual inoculants (*B. subtilis* & *Rhizobium*, *Leguminosarium biovar phaseoli*) also reduced the population of *Meloidogyne sp.*, with the Rhizobium acting as a plant-growth regulator.

Wepukhulu *et al.*[117] found that application of *Bacillus* alone as well as with manure effectively suppressed the population of *Meloidogyne* spp. by 64% and 60%, respectively.

Figure 5 Biocontrol and effect on nematode infection on beans (*Phaseolus vulgaris var.*). (Source: Ayuke, unpublished).

5.4.6 Nematode-destroying Fungi

Nematode-destroying fungi are a group of microfungi that are natural enemies of plant parasitic nematodes.[76,118] They comprise fungi which parasitise nematode eggs and other life stages.[123] Although taxonomically diverse, this group of microorganisms is capable of destroying, by predation or parasitism, microscopic animals such as nematodes, rotifers and protozoans. Collectively, they have the unique ability to capture and infect nematodes in the soil and appear to be widespread in distribution.[76] The actual mechanisms by which the fungi are attracted to the nematodes have not been fully understood. However, it is generally accepted that the nematode cuticle is penetrated, then the nematode is immobilised through infection bulbs, and finally digested by the trophic hyphae produced by the fungus.[120] In some cases, nematode-destroying fungi produce toxins that immobilise or kill nematodes.[121] The group also includes endoparasitic species in such genera as *Harposporium* (see Figure 6), *Nematoctonus*, *Meria* among others, which spend their entire vegetative lives within infected nematodes.[122]

Nematode-destroying fungi have drawn much attention due to their potential as biological control agents of parasitic nematodes of plants and animals.[123,124] Unfortunately, there exist multi-dimensional drawbacks to the realisation of the full potential of the nematode-destroying fungi in the management of parasitic nematodes, especially the phytoparasitic. Lack of reliable methods to visualise the fungi and demonstrate their activity in their natural habitats is a major impediment. Above all, the gaps in knowledge of the ecological factors that influence the occurrence and abundance of nematode-destroying fungi are largely unclear. Due to these factors, this group of fungi has escaped the attention of many scientists, especially in Africa.

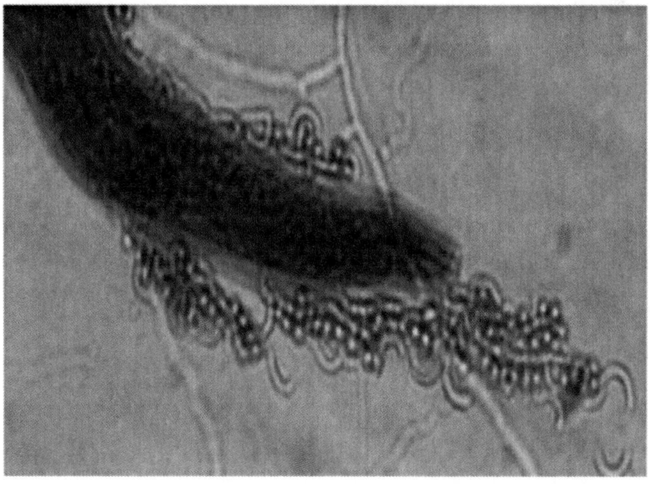

Figure 6 An example of endo-parasitic nematode-destroying fungi: *Harposporium anguillulae* with the conidiaphores and conidia appearing outside the dead nematode. (Source: Wachira[153] (Wachira *et al.*)).

A study on the effect of land use and organic amendments on the occurrence and diversity of nematode-destroying fungi was conducted in Kenya.[153] From the study, it was evident that all the sampled land uses differed in terms of occurrence of nematode-destroying fungi, consistent with previous reports indicating that nematode-destroying fungi were present in all habitats but at different densities and diversities (see Table 7).

Arthrobotrys oligospora was the most abundant species of nematode-destroying fungi in the study area, and this was attributed to the application of inorganic and organic inputs by farmers. Jaffee[119] showed that organic amendments enhanced the build-up of resident nematode-trapping fungi in the soil. Higher soil organic matter content protects plants against nematodes by increasing soil water-holding capacity and enhancing the activity of naturally occurring biological organisms that compete with nematodes in the soil.[126] Apart from the presence of organic matter, the fungi also obtain their carbon and energy from two sources: from organic matter (saprophyte) and from trapping nematodes (parasite), making them adaptable to a wide range of habitats. It is possible that members of the genus were the best adapted to the biotic and abiotic conditions prevailing in the study area. This fungus should be recommended for further study with the aim of developing it as a biological control agent. Such a study should be geared towards growth parameters of the fungus, since biological, chemical and physical factors of the soil are known to inhibit fungal growth by fungistatic compounds and is made even more complicated by crop rotations. The ability of this fungus as a biological control agent could be improved through genetic engineering and then packaged for biological control purposes. Apart from introduction of particular species from the genus, agricultural practices that stimulate build-up of the fungi could be identified and recommended for adoption by farmers. The study also revealed that increased land-use intensity resulted in increased occurrence and diversity of nematode-destroying fungi. This, however, was contrary to the expectation that beneficial microorganisms decrease with increased intensity of land use.[125] A number of explanations were used to account for the higher frequency of occurrence of nematode-destroying fungi

Table 7 Effect of land use on frequency of isolation, richness and diversity of nematode-destroying fungi in Taita Taveta district, Kenya. (Source: Wachira *et al.*).[153]

Land use	*Frequency of isolation %*	*Mean evenness*	*Mean richness*	*Mean Shannon*
Forest	5.8	0.375	0.625	0.17
Maize/bean	27.9	1.000	3.000	1.07
Napier	20.9	1.000	2.250	0.76
Shrub	11.6	0.625	1.250	0.36
Vegetables	33.7	1.000	3.625	1.26
P-value	3.81×10^{-07}	7.139×10^{-05}	3.81×10^{-07}	1.062×10^{-06}

in the habitats that are subject to regular disturbance compared to the stable ecosystems like shrub land and indigenous forest. It was also possible that fungal tissues were fragmented and scattered in the course of farm operations, thus increasing their frequency of detection. As such, agricultural practices can exert positive or negative impacts on other microorganisms in the soil.[127] According to Wang *et al.*,[128] some agricultural inputs stimulate build-up of nematode-trapping fungi, hence the observed diversity, evenness and richness with increased land-use intensity compared to land uses such as forest or shrub land which are materially unchanged by human activity. Intensive cultivation is characterised by increased movement of soil, which may result in increased spread of the microorganisms in the field. Soil disturbance, coupled with frequent changes in crop cover, subjects the soil biota to stress, making it difficult for a particular species to establish itself in the soil to out-compete the others. In contrast, soils under forest and shrub are less disturbed, meaning that certain species of nematode-destroying fungi are able to establish and suppress other species that are poorly suited to compete effectively.

5.5 Farming Systems and Soil-borne Pests and Diseases

In conventional agriculture, addition of lime, inorganic fertilisers and pesticides can change the physical and chemical nature of the soil environment, thereby altering the number of organisms and the ratio of different groups of organisms, resulting in adverse effects characterised in part by an increase in soil-borne pests and diseases. Soil-borne pathogens (such as plant parasitic nematodes, fungi, bacteria, phytoplasma, protozoa and viruses) are among the most underestimated of the factors which affect plant productivity in tropical regions. The reasons for the greater severity of soil-borne diseases and pests in the tropics are the generally favourable climatic conditions, the greater pathogenicity of pest species and the more severe disease complexes.[129] In addition, cropping systems in tropical regions are generally more diverse and less reliant on chemical inputs compared to those in temperate regions. There is also a greater diversity of nematodes and other pests in tropical regions.[67] Table 8 lists some of the most common soil-borne pathogens in the tropics and the crops and trees that may be affected in different systems.

In general, plants infected by soil-borne pathogens suffer from root rot, collar rot, root blackening, wilting, stunting or seedling damping-off diseases. To some extent, losses associated with soil-borne pathogens may be reduced by a 4–5 year crop rotation programme, but this is not feasible due to economic reasons. One way in which the soil-borne pathogens can be indirectly suppressed is through the incorporation of organic amendment matter to mineral soils. In addition to improving tilth, aeration and drainage of soils where organic matter is incorporated, additional benefits occur such as proliferation of populations of beneficial soil microorganisms. This was demonstrated by Langat *et al.*[130] where amending soils with organic substrates including baggase, molasses, tea and flower composts contributed to a change

Table 8 Common soil-borne pathogens on major field crops in the tropics.

Pest/Pathogen	*Diseases*	*Common host crops*	*Reference*
Fungi			
Fusarium spp.	Wilt, crow rot, blackleg	Vegetables, banana, bean, coffee, cotton, melon, potato, tomato, cowpea, *Crotalaria* spp., *Sesbania* spp.	6, 154
Phytophthora spp.	Root rots, blights	Vegetables, soybean, cowpea, cocoa, citrus, tobacco	155
Pythium spp.	Damping off diseases	Vegetables, soybean, cowpea, common bean, chick pea	156
Rhizoctonia spp	Root rots, blights	Vegetables, soybean, cowpea, common bean, chick pea	156, 157
Sclerotium spp.	Collar rot, southern blight	Solanaceous crops, root and tuber crops, legumes, rice, *Mucuna* spp., Sesbania spp.	156, 158
Macrophomina phaseolina	Black root rot	Soybean	157
Bacteria			
Ralstonia solanacearum	Bacterial wilt	Tomato, pepper, eggplant, groundnut	
Xanthomonas campestris	Black rot	Kale, cabbage, broccoli	159
Agrobacterium tumefaciens	Crown gall	Roses, grape vines, stone fruit trees	
Nematodes			
Meloidogyne spp. (root-knot nematodes)	Root knot disease	Vegetables, legumes, tubers, coffee, *Sesbania* spp., *Tephrosia* spp.	160
Pratylenchus spp. (lesion nematodes)	Root lesion disease	Cereal crops, root and tuber crops, banana, coffee, tea, *Arachis* spp., forage grasses *Crotalaria* spp., *Senna* spp.	160
Radopholus similis (burrowing nematodes)		Banana, citrus, pepper and palms	160

in the nematode community structure by significantly increasing the abundance of beneficial nematodes in the soil. An important consideration is that all soils have an inherent natural level of disease-suppressive activity. In most soils, long-term management, or lack thereof, can either reduce or increase this level of suppression. A number of land-management factors such as intensification in cropping, amending soils with organic matter, weed management, and stubble retention have been shown to increase soil suppressiveness to cereal root disease. The concept of a 'suppressive soil' was first described by Menzies[131] to explain the phenomenon of soils that suppressed Streptomyces potato scab. To date, natural suppressive soils have been described containing a number of soil-borne pathogens such as *Gaeumannomyces graminis* var. tritici (take-all disease of wheat), *Fusarium oxysporum* (wilt diseases of tomato, radish, banana and others), *Phytophthora cinnamon* (root rot of eucalyptus), *Pythium* spp. and *Rhizoctonia solani* (damping-off of seedlings of several crops, including sugar beet and radish), *Thielaviopsis basicola* (black root rot of tobacco, bean, cherry trees and others), Streptomyces scabies (bacterial potato scab; that is, lesions on potato tubers), *Ralstonia solanacearum* (bacterial wilt of tomato, tobacco and others), and *Meloidogyne incognita* (root swelling and root-knot galls caused by this nematode on several crops, mostly in tropical and subtropical countries).

6 Mitigation of Soil Degradation through Integrated Soil Fertility Management (ISFM) Approaches: Sustainable Soil-management Practices/Systems

Crop yields in large parts of sub-Saharan Africa are low due to declining soil fertility associated with continuous cropping and sub-optimal fertiliser use. With the liberalisation of trade and introduction of structural adjustment programmes, fertiliser costs have increased and most small-scale farmers can no longer afford them, while the challenge of increasing and maintaining crop yields to sustain the growing population in most countries south of the Sahara has remained. Animal manure, as an alternative for maintaining soil fertility and crop productivity, is available in inadequate amounts and is of low quality due to poor handling and poor quality livestock feeds.[132,133] Technologies such as improved fallow systems[49] and use of organic inputs[134,135] have been demonstrated to increase crop yields, but often organic resources used alone provide insufficient nutrients to build up longer-term soil fertility and sustain crop yields.[136] Integrated Soil Fertility Management (ISFM), *i.e.* combined use of organic and inorganic fertilisers, has been recommended for increasing nutrient use efficiency (NUE) among farmers in SSA.[52,136] One of the major challenges in such low-input systems is to develop ways of managing organic matter to optimise the maintenance of SOM, improve soil structure and enhance water- and nutrient-use efficiencies. One aspect of ISFM that is often ignored is that it offers perspective for the manipulation of community composition and activities of soil biota through the judicious management of organic inputs. Especially the stimulation of earthworm and termite activity may contribute to decomposition and humification of organic residues, maintenance of soil structure and aggregate stability, and overall restoration of degraded soils.[66] In a wider sense, the elucidation of biodiversity of soil organisms has high priority in global biodiversity research, as it appears to be key to understanding their role in soil ecosystem processes and services.[137,138]

7 Biodiversity of Tropical Soils: Socioeconomic, Institutional and Policy Issues

Conservation of natural resources, including tropical soil biodiversity, has remained one of the most challenging problems, partly due to declining fertility of tropical soils; hence the reduced capacity of such soils to produce adequate food to meet household food requirements.[139] The ensuing pursuit for household food security has, on the other hand, tended to encourage adoption of practices that degrade soils. Generally, soil degradation gradually diminishes the capacity of individual farmers and communities to raise sufficient incomes from farming activities which, in turn, results in the inability to undertake critical investments needed to conserve the soil and preserve biodiversity. It also diminishes opportunities for such households to satisfy their nutritional needs. At the same time, the households become vulnerable to external shocks and often disinvest in critical productive assets to cope with such shocks.[140] Thus, degradation of natural resources including land (and soil biodiversity) has the effect of entrenching nutritional and asset poverty, which in turn reinforce natural resource degradation, thus creating a vicious circle. This nexus between worsening poverty and degradation of natural resources raises fundamental questions of the best strategies for managing soil biodiversity in the tropics. These challenges are highest in many developing regions, representing the intersection of hot-spots of widespread poverty and fragile ecosystems (*e.g.* arid and semi-arid areas, highland regions).[139,141]

Governments, donors and development partners in many developing countries have devoted substantial resources to developing and promoting a diverse mix of sustainable soil conservation practices. The technologies promoted in this mix have included indigenous and introduced structural technologies and agronomic practices, usually aimed at enhancing soil productivity. Some of the structural methods include soil and stone bunding and terracing, while the agronomic practices include minimum tillage, organic and inorganic fertilisers, pesticides, grass strips, and agro-forestry techniques. In addition, a number of agro-forestry technologies, in particular alley cropping, have been promoted mainly because of their ability enhance soil organic matter and, in cases involving leguminous plants, replenish soil nitrogen through nitrogen fixation.[142]

Despite the increasing efforts made and the growing policy interest, there has been limited focus on the promotion of soil biodiversity, especially below-ground biodiversity, in the tropics. Instead, farm households have increased the use of soil fertility management and agronomic practices that are usually promoted to enhance agricultural productivity but tend to hurt the below-ground micro- and macro-organisms. This section first reviews the soil conservation approaches pursued in the past followed by a discussion of socioeconomic (*e.g.* incentives and capacity) and institutional (including market access and policy) and information-related factors that condition the adoption of sustainable soil conservation practices likely to affect the biodiversity of tropical soils.

7.1 Approaches to Soil Conservation: A Historical Perspective

In order to stimulate widespread adoption and adaptation of soil conservation practices in tropical agriculture, especially in marginal and vulnerable environments, three major approaches have been used,[143] namely, top-down interventions, populist or farmer-first, and neo-liberal approaches. The early soil conservation approaches used the top-down approach to promoting the use of conservation practices. The practices promoted mainly involved structural methods used to prevent soil erosion. The approach earned its name from the lack of farmer participation in technology design and the use of command-and-control type policies used in implementation of the externally developed structural measures. The policies pursued under this approach included forced adoption of soil erosion control and planting of trees on hillsides, both of which have the potential to improve soil biodiversity by either retaining or replenishing the soil organic matter. However, the policies were largely driven by fear of future consequences of inaction. Nonetheless, this approach to soil conservation continued in several tropical areas (especially in Africa) until the mid 1980s.[144,145] The majority, however, failed to realise expected gains due to lack of incentive and initiative by households, resulting in the abandonment of the technologies as soon as the authorities were not involved.

The experiences gained from the failure of top-down policies were used to formulate a new approach referred to as the "populist" approach. This approach made the farmer central to program design and implementation of soil conservation activities. It had its foundations in the book *Farmer First*.[146] The approach stressed a small-scale and bottom-up participatory approach to soil conservation using homegrown technologies[147] and rejected wholesale technology transfer. However, it faced difficulties because of its failure to address the economic, institutional and policy environments in which farmers operate.[143,148] Consequently, development agencies developed the third approach, namely the neo-liberal approach, which advocated the need to understand the structure of incentives that impede the use of soil conservation technologies. The neo-liberal approach recognised the essential role of farmer innovation but emphasised the critical role of markets, policies and institutions in stimulating and inducing farmer innovation, adoption and adaptation of suitable soil conservation options.[139] It especially focused on making soil conservation attractive and economically rewarding to farmers. The approach spearheaded the adoption of productive technologies and improved access to markets, which usually spur farmer investments in sustainable soil conservation options due to increased agricultural revenues.

The approach used in promoting soil conservation in agriculture has further changed in the last few years, moving instead towards the concept of sustainable land management (SLM) both at the farm and landscape level.[148] While there is no single all-encompassing definition of SLM, it has been suggested[149] that SLM implies a system of technologies that aims to integrate ecological, socio-economic and political principles in the management of land for agricultural and other purposes to achieve intra- and inter-generational equity. This broadening of the concept of soil conservation shows the complexity of the challenges it entails. The following section examines these challenges in the context of incentives and capacity variables, the institutional and the information-related factors that condition adoption.

7.2 Drivers of Farmers' Use of Sustainable Soil Conservation Practices

Farmers adopt new practices that enhance soil biodiversity only when the switch from the old to new methods offers additional gains either in terms of higher net returns, lower risks, or both. Thus farmers are likely to adopt soil biodiversity-enhancing practices only when the additional benefits from such investments outweigh the added costs.[150] Investment in such soil conservation practices is often just one of the many investment options available to farmers. They can therefore defer undertaking such conservation investments until the gains from such investments are perceived to be at least equal to the next-best investment opportunities available to them.[151] That is, farmers will implicitly compare the expected costs and benefits and then invest in options that offer highest net returns in terms of income or reduced risk. This implies that, in cases where private costs of investment in soil biodiversity outweigh the benefits, voluntary adoption will be greatly hampered and may only occur if the society is willing to internalise some of the costs by offering subsidies to farmers. This is indeed the reason why some development experts promote the payment for environmental services.[151]

The literature identifies a number of factors that condition the adoption of soil conservation practices in agriculture. These factors relate to incentives the farmers have and the capacity of such farmers to adopt better practices. Farmers can be constrained to adopt otherwise profitable (or economically attractive) interventions due to asset poverty (*i.e.* low endowment with needed capital items), imperfect information, poorly functioning markets, bad policies, and institutional factors. Thus the factors that condition the adoption of soil biodiversity can be broadly categorised into incentive factors, capacity factors, institutional (*e.g.* markets and policy) factors and information-related factors.

8 Synthesis

In summary, the recognised form of land degradation affecting the major soil types in sub-Saharan Africa are erosion, physical and chemical degradation, which includes salinisation, sodification, acidification and the depletion of plant nutrient content in the soil. Biological degradation is also a major contributor leading to loss of soil organic matter and soil biodiversity. All these forms of degradation lead to a lowering of soil fertility and land productivity. Land degradation problems are now recognised as being one of the major contributors to the persistent food deficit and high poverty levels in the sub-Saharan Africa. The main causes of low land productivity in smallholder farms include very low use of organic and inorganic fertilisers; poor tillage practices, especially for hard setting soils such as Luvisols, Lixisols and Acrisols; excessive soil erosion by water and wind, affecting almost all the major soil types; lack of attention to soil acidity for soils with acidity problem; poor conservation and management of rain water for enhanced soil moisture conservation on soils occurring in rolling to undulating topography; and poor land-use planning. Concerted and well-planned action therefore needs to be taken to build soil fertility and minimise land degradation on small-scale farms. Some of the important action points are:

- Developing well-defined and specific activities to enhance plant nutrient levels as a long–term programme through consistent use of both organic and inorganic fertilisers. According to World Bank figures, Africa uses only 14 kg of fertiliser per hectare compared with 1150–2000 kg in East Asia and Europe. Use of both organic and inorganic fertilisers have resulted in improved soil physical and chemical properties and increased crop yields for some highly weathered tropical soils.
- Giving adequate attention to the problem of soil acidity and finding better ways of promoting plant nutrient availability and uptake.
- Developing and adapting suitable rotations using legumes and green manure.
- Promoting agroforestry and farm forestry for better soil fertility and increased land productivity to answer multiple needs at the farm level and beyond.
- Creating programmes to deal with the issues of tillage and depth of root bed to create sufficient storage capacity for plant nutrients and water, especially for soils with a compacted sub-soil. Further issues of the energy required and the development of new or improved tillage systems and equipment need to be dealt with as crucial elements in the process. Such improved methods of tillage should lessen the problem of hardpans and plough soles. This will greatly enhance soil water uptake for plant growth.
- Developing efficient systems of irrigation that increase production without degrading the soil.
- Adopting soil conservation measures that are simple, effective and affordable.
- Within the context of Integrated Soil Fertility Management (ISFM), soil biota are responsible for the key ecosystem functions of decomposition and nutrient cycling, soil organic matter synthesis and mineralisation, soil structural modification and aggregate stabilisation, nitrogen fixation, nutrient acquisition, regulation of atmospheric composition, the production of plant growth substances and the biological control of soil-borne pests and diseases Understanding biological processes is not as well advanced as those that are related to soil physical and chemical properties, creating opportunities for breakthroughs in biotic function to better service agriculture.

To summarise, understanding the soil is the key to its improvement, as there are many physical, chemical and biological properties of the various soil types that affect plant growth.

References

1. FAO, The conservation of Lands in Asia and Pacific, Natural Resources Management and Environment Department, 2008.

2. J. S. C. Mbagwu, Aggregate Stability and Soil Degradation in the Tropics, *Department of Soil Science*, University of Nigeria, Nsukka, 2003, pp. 1–7.
3. E. M. A. Smaling, PhD thesis, Agricultural University, Wageningen, The Netherlands, 1993.
4. J. Van der Pol, *Bull., Dep. Agric. Res. R. Trop. Inst., Amsterdam* , 1992, 325.
5. C. K. K. Gachene, N. J. Jarvis, H. Linner and J. P. Mbuvi, *Soil Sci. Soc. Am. J.*, 1997, **61**, 559–564.
6. E. K. Kiprop, PhD thesis, University of Nairobi, 2001, pp. 218.
7. P. A. Sanchez, C. A. Palm, L. T. Szott, E. Cuevas and R. Lal, Dynamics of Soil Organic Matter, in *Tropical Ecosystems*, ed. D. C. Coleman, J. M. Oades, G. Uehara, NifTAL Project, University of Hawaii, Honolulu, HI, 1989, pp. 125–152.
8. R. Lal and B. A. Stewart, *Advances in Soil Science*, vol. 11, Springer-Verlag, Berlin and Heidelberg, 1990, pp. 345.
9. C. K. K. Gachene and G. Kimaru, *RELMA Technical Handbook Series 30*, RELMA/Sida, Nairobi, 2003, pp. 146.
10. A. Bationo, B. Waswa, J. M. Okeyo, F. Maina and J. Kihara, *Exploring Scientific Facts*, Springer, Dordretch Heidelberg London New York, 2011, pp. 1363.
11. M. Eilitta, J. Mureithi and R. Derpsch, *Experiences from Tropical and Subtropical Regions*, Kluwer Academic Publishers, Dordretch, The Netherlands, 2004, pp. 361.
12. V. Rutunga, K. G. Steiner, N. K. Karanja, C. K. K Gachene and G. Nzabonihankuye, *Biotechnol. Agron. Soc. Environ.*, 1998, **2**(2), 35–142.
13. ACT (African Conservation Tillage Network), *Proceedings of the Third World Congress on Conservation Agriculture*, 3–7 October, 2005, Nairobi. ACT Network, Nairobi, 2008.
14. V. I. Sijali, Drip irrigation: options for smallholder farmers in eastern and southern Africa, *RELMA/ Sida*, Nairobi, 2001, pp. 60.
15. D. Thomas, *Ministry of Agriculture*. English Press, Nairobi, Kenya, 1997, pp 296.
16. F. M. S. Moreira, J. E. Huising, D. Bignell, A handbook of Tropical Soil Biology: Sampling and characterization of Below-ground Biodiversity, Earthscan, Publishers, UK, 2008.
17. P. L. Woomer, A. Martin, A. Albretch, D. V. S. Resck and H. W. Scharpenseel, *The Biological Management of Tropical Soil Fertility*, ed. P. L. Woomer and M. J. Swift, John Wiley and Sons, Chichester, 1994, pp. 47–51.
18. N. Uphoff, A. Ball, E. Fernandez, H. Herren, O. Husson, M. Laing, C. Palm, J. Pretty and P. A. Sanchez, Biological Approaches to Sustainable Soil Systems, CRC Press, Boca Raton, USA, 2006, pp. 784.
19. L. Brussaard and N. G. Juma, *Humic Substances in Terrestrial Ecosystems*, ed. A. Piccolo, Elsevier, Amsterdam, 1996, pp. 329–359.
20. R. Lal, *Agric. Ecosyst. Environ*, **24**, 101–116.

21. M. J. Swift, J. Vandermeer, P. S. Ramakrishnan, J. M. Anderson, C. K. Ong and B. A. Hawkins, *Functional Roles of Biodiversity: A Global Perspective*, ed. H. A. Mooney, J. H. Cushman, E. Medina, O. E. Sala and E. D. Schulze, John Wiley and Sons Ltd, New York, USA, 1996, pp. 261–298.
22. P. Lavelle, *International Workshop of Tropical Soil Biology: Opportunities and Challenges for an African Agriculture*, March, 16–19, 1999, Nairobi, Kenya.
23. H. Bossuyt, J. Six and P. F. Hendrix, *Eur. J. Soil Sci.*, 2004, **55**, 393–399.
24. P. Lavelle, L. Brussaard and P. Hendrix, Earthworm management in tropical agroecosystems, CABI, Wallingford, United Kingdom, 1999, pp. 300.
25. M. M. Pulleman, J. Six, A. Uyl, J. C. Y. Marinissen and A. G. Jongman, *Appl. Soil Ecol.*, 2005, **29**, 1–15.
26. M. M. Pulleman, J. Six, N. van Breemen and A. G. Jongman, *Eur. J. Soil Sci.*, 2004, **10**, 1–15.
27. E. Blanchart, A. Albrecht, G. Brown, T. Decaens, A. Duboisset, P. Lavelle, L. Mariani and E. Roose, *Agric. Ecosyst. Environ.*, 2004, **104**, 303–315.
28. B. M. Doube, M. H. Ryder, C. W. Davoren and P. M. Stephen, *Biol. Fertil. Soils*, 1994, **18**, 169–174.
29. K. M. Harinikumar and D. J. Bagyaraj, *Biol. Fertil. Soils*, 1994, **18**, 115–118.
30. M. Szczech, W. Rondomanski, M.W. Brzeski, U. Smolinska and J. F. Kotowski, *Biol. Agric. Hortic.*, 1993, **10**, 47–52.
31. D. O. Asawalam, V. E. Osodeke, O. J. Kamalu, and I. K. Ugwa, *Commun. Soil Sci. Plant Anal.*, 1999, **30**(11–12), 1691–1696.
32. C. Kooyman and R. F. M, *Wageningen Agicultural University Papers*, Wageningen, The Netherlands, 1987, 87–3, pp. 120.
33. A. Mando, L. Brussaard and L. Stroosnijder, *Restorat. Ecol.*, 1999, **7**, 33–41.
34. M. Sarr, C. Agbogba, A. Russell-Smith and D. Masse, *Appl. Soil Ecol.*, 2001, **16,** 283–290.
35. W. G. Wielemaker, Soil formation by termites, a study in the Kisii area, Kenya, Wageningen University, Wageningen, The Netherlands, 1984, pp. 146.
36. A. Mando, L. Stroosnijder and L. Brussaard, *Geoderma*, 1996, **74**, 107–113.
37. A. Mando, *Land Degradat. Dev*. 1997a, **8**, 269–278.
38. A. Mando, *Soil Technol.*, 1997b, **11**, 121–138.
39. E. Ou'edraogo, A. Mando and L. Brussaard, *Eur. J. Soil Biol.*, 2006, **42**, 275–277.
40. P. Jouquet, L. Mamou, M. Lepage and B. Velde, *Eur. J. Soil Sci.*, 2002, **53**, 521–527.
41. P. D. Jungerius, J. A. M. van den Ancker and H. J. Mücher, *Catena*, 1999, **34**, 349–363.
42. A. Mando and L. Brussaard, *Biol. Fertil. Soils*, 1999, **29**, 332–334.
43. C. Martius, *Pedobiologia*, 1994, **38**, 407–428.

44. N. M. Collins, *Tropical Rainforests: Ecology and Management*, ed. S. L. Sutton, Blackwell, Oxford, 1983.
45. N. M. Collins, Oecologia, 1981, **51**, 389–399.
46. D. Lopez-Hernandez, *Soil Biol. Chem.*, 2001, **33**, 747–753.
47. K. W. Nyamapfene, *Trop. Agric. (Trinidad)*, 1986, **63**, 191–192.
48. B. M. Sekamatte, M. Ogenga-Latigo and A. Russell-Smith, *Crop Protect.*, 2003, **22**, 87–93.
49. G. Sileshi, P. L. Mafongoya, F. Kwesiga and P. Nkunika, *Agric. Forest Entomol.*, 2005, **7**, 61–69.
50. P. Chivenge, B. Vanlauwe, R. Gentile, R. Wangechi, D. Mugendi, C. Van Kessel and J. Six, *Agron. J.*, 2009, **101**, 1266–1275.
51. R. Gentile, B. Vanlauwe, C. Van Kessel and J. Six, *Agric. Ecosyst. Environ.*, 2009, **131**, 308–314.
52. B. Vanlauwe, A. Bationo, J. Chianu, K. E. Giller, R. Merckx, U. Mokwunye, O. Ohiokpehai, P. Pypers, R. Tabo, K. Shepherd, E. Smaling, P. L. Woomer and N. Sanginga, *Outlook Agric.*, 2011, **39**(1), 17–24.
53. R. Dudal and J. Deckers, *Soil Organic Matter Dynamics and Sustainability of Tropical Agriculture*, ed. K. Mulongoy, R. Merckx, Wiley-Sayce co-publication, West Sussex, United Kingdom, 1993, pp. 377–388.
54. M. J. Swift and P. L. Woomer, *Dynamics of Soil Organic Matter in Relation to the Sustainability of Agricultural Systems,* ed. R. Merckx and K. Mulongoy, John Wiley, Chichester, 1992, pp. 3–18.
55. P. L. Woomer and M. J. Swift, *The Biological Management of Tropical Soil Fertility*, John Wiley and Sons, Chichester, 1994, pp. 243.
56. F. O. Ayuke, L. Brussaard, B. Vanlauwe, J. Six, D. K. Lelei, C. Kibunja and M. M. Pulleman, *Appl. Soil Ecol.*, 2011a, **48**, 53–62.
57. F. O. Ayuke, M. M. Pulleman, B. Vanlauwe, R. G. M. de Goede, J. Six, C. Csuzdi and L. Brussaard, *Agric. Ecosyst . Environ.*, 2011b, **148**, 148–154.
58. J. P. Curry, *Earthworm Ecology*, ed., C. A. Edwards, CRC Press, Boca Raton, London, New York, 2nd edn, 2004, pp. 91–111.
59. P. A. Sanchez, *Science*, 2002, **295**, 2019–20.
60. P. A. Sanchez, K. D. Shepherd, M. J. Soule, F. M. Place, R. J. Buresh, N. I. Anne-Marie, A. U. Mokwunye, F. R. Kwesiga, C. G. Ndiritu and P. L. Woomer, *Replenishing Soil Fertility in Africa*, ed. R. J. Buresh, P. A. Sanchez and F. Calhoun, Soil Science Society of America (SSSA). Special publication No 51, USA, 1997.
61. J. Six, E. T. Elliot and K. Paustian, *Soil Biol. Biochem.*, 2000, **32**, 2099–2103.
62. J. Six, E. T. Elliot and K. Paustian Soil Sci. *Soc. Am. J.*, 1999, **63**, 1350–1358.
63. E. Ou'edraogo, *Tropical Resource Management Papers*, 51, Wageningen University, 2004, pp. 193.
64. A. D. Mackay and E. J. Kladivko, *Soil Biol. Biochem.*, 1985, **17**, 851–857.
65. W. Tinzara and E. Tukahirwa, *Afnet V Report*, 1995, pp. 23.
66. L. Brussaard, M. M. Pulleman, E. Ou′edraogo, A. Mando and J. Six, *Pedobiologia*, 2007, **50**, 447–462.

67. M. H. Beare, M. V. Reddy, G. Tian and S. C. Scrivasta, *Appl. Soil Ecol.*, 1997, **6**, 87–108.
68. J. M. Dangerfield, *Report on Characterization of an Experimental Field in KARI Farm, Muguga, Kenya*, ed. M. R. Rao and R. J. Scholes, ICRAF, Nairobi, Kenya, 1993, pp. 51–67.
69. F. O. Ayuke, M. R. Rao, M. J. Swift and M. L. Opondo-Mbai, *East African Agric. Forest J.*, 2003, **69**(1–2), 131–137.
70. S. J. Fonte, T. Winsome and J. Six, *Appl. Soil Ecol.*, 2009, **41**, 206–214.
71. G. Tian, B. T. Kang and L. Brussaard, *Soil Biol. Biochem.*, 1997, **29**, 369–373.
72. F. O. Ayuke, MPhil thesis of Moi University, Eldoret, Kenya, 2000.
73. H. Bossuyt, J. Six and P. F. Hendrix, *Soil Biol. Biochem*, 2005, **37**, 251–258.
74. J. C. Moore, *Agric. Ecosyst. Environ.*, 1994, **51**, 239–247.
75. C. W. Maribie, G. H. N. Nyamasyo, P. N. Ndegwa, J. K. Mung'atu, J. Lagerlof and M. Gikungu, *Trop. Subtrop. Agroecosyst.*, 2011, **13**, 11–26.
76. N. H. Birgit, B. J. Hans and T. Anders, *Encyclopedia of Life Sciences*, Johny Wiley and Sons Ltd., New York, USA, 2006, pp. 1–11.
77. J. J. Muturi, J. P. Mbugi, J. M. Mueke, J. Lagerlof, J. K. Mung'atu, G. H. N. Nyamasyo and M. Gikungu, *Trop. Subtrop. Agroecosyst.*, 2011, **13**, 35–42.
78. R. C. Burns and R. F. W Hardy, Nitrogen fixation in Bacteria and Higher plants, Academic Press Inc., New York, USA, 1975, pp. 39–55.
79. E. Martinez–Romero, L. Segovia, F. M. Mercante, A. A. Franco, P. Graham and M. A. Pardo, *Int. J. Syst. Bacteriol,*. 1991, **48**, 107–119.
80. J. P. H. Kahindi and N. K. Karanja, *Encyclopedia of Life Support Systems*, EOLSS Publishers Co. Ltd. Oxford, UK, 2006; http://www.oels.net
81. A. Sy, *J. Bacteriol.*, 2001, **183**, 214–220.
82. R. T. Samba, P. Lajudie, M. Gillis, M. Neyra, M. M. S Barelto and B. Dreyfus, *Symbiosis*, 1999, **27**, 259–268.
83. D. W. Odee, K. Haukka, S. G. McInroy, J. I. Sprent, J. M. Sutherland and J. P. W. Young, *Soil Biol. Biochem.*, 2002, **34**, 801–811.
84. B. Anyango, K. J. Wilson, J. L. Beynon and K. E. Giller, *Appl.Microbiol.*, 1995, **61**, 4016–4021.
85. G. M. Mwenda, N. K. Karanja, H. M. Boga, J. H. P. Kahindi, A. Muigai, D. Odee, *Trop. Subtrop. Agroecosyst.*, 2011, **13**, 1–10.
86. S. M. Mwangi, N. K. Karanja, H. M. Boga, J. H. P. Kahindi, A. Muigai, D. Odee, *Trop. Subtrop. Agroecosyst.*, 2011, **13**, 109–118.
87. N. K. Karanja, P. L. Woomer and S. Wangaruro, *Microbial Diversity and Ecosystem Function*, ed. D. Allsopp, D. L. Harksworth and R. R. Colwell, CABI, 1995, pp. 447–453.
88. A. Bationo, B. Waswa, J. M. Okeyo, F. Maina, J. Kihara and U. Mokwunye, Fighting poverty in Sub-Saharan Africa: The multiple Roles of Legumes in Integrated Soil Fertility Management, Springer Publishers, Springer Dordrecht Heidelberg London New York, 2011, pp. 246.
89. J. L. Harley and S. E. Smith, Mycorrhizal symbiosis, Academic, London, UK, United Kingdom, 1983, pp. 483.

90. N. Sanginga and P. L. Woomer, Integrated Soil Fertility Management in Africa: Principles, Practices and Developmental Process, Tropical Soil Biology and Fertility Institute of the International Centre for Tropical Agriculture, Nairobi, Kenya, 2009, pp. 263.
91. H. Tomlison, Z. Taklehaimanot, A. Traore and E. Olapade, *Agroforestry Syst.*, 1995, **19**, 241–252.
92. K. D. Shepherd, J. Jefwa, J. Wilson, J. Ndufa, K. Ingleby and K. W. Mbuthia, *Biol. Fertil. Soils*, 1996, **22**, 16–21.
93. J. M. Jefwa, J. Mung'atu, P. Okoth, E. Muya, H. Roimen and S. Njuguini, *Trop. Subtrop. Agroecosyst.*, 2009, **11**, 277–290.
94. K. M. Cooper and P. B. Tinker, *New Phytologist*, 1978, **81**, 43–52.
95. D. J. Bagyaraj and A. Varma, *Adv. Microbial Ecol.*, 1995, **14**, 119–142.
96. E. Sieverding, Vesicular-arbuscular mycorrhiza management in tropical agro-ecosystems, *German Technical Cooperation* (GTZ), Eschborn, Germany, pp. 371. ISBN 3 88085 462 9.
97. M. C. Brundrett, *Adv. Ecol. Res.*, 1991, **21**, 171–213.
98. B. R. Glick, *Can. J. Microbiol.*, 1995, **41**, 109–117.
99. J. C. Biswas, J. K. Ladha and F. B. Dazzo, *Soil Sci. Soc. Am. J.*, 2000, **64**, 1644–1650.
100. B. R. Glick, D. M. Penrose and J. Li, *J. Theor. Biol.*, 1998, **190**, 63–68.
101. E. S. Paterno, Plant growth promoting rhizobacteria: their potential in improving crop productivity, UP Los Banos, Laguna, Philippines, 1997.
102. A. Karawal, *J. Plant Pathology*, 2009, **91**, 61–63.
103. Y. Eland, *Crop Protect.*, 2000, **19**, 709–714.
104. L. G. Hjeljord and A. Tronsmo, *Trichoderma* and *Gliocladium,* ed. C. P. Kubicek and G. E. Harman, Taylor & Francis, Ltd., London, UK, 1998, pp. 135–151.
105. C. R. Howell, L. E. Hanson, R. D. Stipanovic, L. S. Puckhaber, *Phytopathology*, 2000, **90**, 248–252.
106. I. Yedidia, N. Benhamou and I. Chet, *Appl. Environ. Microbiol.*, 1999, **65**, 1061–1070.
107. S. Okoth, J. A. Otadoh and J. O. Ochanda, *Trop. Subtrop. Agroecosyst.*, 2011, **13**, 65–71.
108. S. Okoth, P. Okoth, P. M. Wachira and H. Roimen, *Trop. Subtrop. Agroecosyst.*, 2009, **11**, 291–302.
109. Y. Elad, I. Chet and Y. Katan, *Phytopathology*, 1980, **70**, 119–121.
110. H. Basim, S. B. Ozturk and O. Yegen, presented at the GAP-Environmental Symposium, Sanliurfa, Turkey, 1999, pp. 137–144.
111. C. Henis, *Current Perspectives in Microbial Ecology,* ed. M. J. Klug and C. A. Reddy, American Society of Microbiology, Washington, 1984, pp. 353–361.
112. I. Chet and J. Inbar, *Appl. Biochem. Biotech.*, 1994, **48**, 37–43.
113. R. J. Cook and K. F. Baker, *Am. Phytopathol.. Soc.*, St Paul, MN, 1983, pp. 539.

114. E. N. Siameto, S. Okoth, N. O. Amugune and N. C. Chege, *Trop. Subtrop. Agroecosyst.*, 2009, **13**, 81–90.
115. J. W. Kimenju, A. M. Kagundu, J. H. Nderitu, F. Mambala, G. K. Mutua and G. M. Kariuki, *Asian J. Plant Sci.*, 2008, **7**, 404–408.
116. N. K. Karanja, G. K. Mutua and J. W. Kimenju, *J. Trop. Subtrop. Agroecosyst.*, 2007, **13**, 27–34.
117. M. Wepukhulu, J. Kimenju, B. Anyango, P. Wachira and G. Kyallo, *Trop. Subtrop. Agroecosyst.*, 2011, **13**, 27–34.
118. Y. Yang, E. Yang, Z. An and X. Liu, *Proc. Natl. Acad. Sci. U. S. A.*, 2007, **104**, 8379 – 8384.
119. B. A. Jaffee, *J. Nematol.*, 2004, **36**, 267–275.
120. J. J. Bordallo, L.V. Lopez-Liorca, H. B. Jansson, J. Salines, L. Persmark and L. Asonsio, *New Phytopathol.*, 2002, **154**, 491–499.
121. J. V. Araújo, M. A. Stephano and W. M. Sampaio, *Vet. Arh.*, 1999, **69**, 69–78.
122. L. Timm, D. Pearson and B. Jaffee, *Mycologia*, 2001, **93**, 25–29.
123. H. Jansson and C. Persson, *Mycologia*, 2000, **92**, 10–15.
124. S. G. Masoomeh, R. A. Mehdi, R. B. Shahrokh, E. Ali, Z. Rasoul and E. Majid, *Iran. Biomed. J.*, 2004, **8**, 135–142.
125. J. Vandermeer, M. van Noordwijk, J. M. Anderson, C. Ong, and I. Perfecto, *Agric. Ecosyst. Environ.*, 1998, **67**, 1–22.
126. G. Kaskavalci, *Turk. J. Agric.*, 2007, **31**, 159–167.
127. A. Akhtar and A. Malik, *Bioresource Technol.*, 2000, **74**, 35–47.
128. H. Wang, D. Sipes and P. Schmitt, *Agric. Ecosyst. Environ.*, 2003, **94**, 197–203.
129. W. F. Mai and G. S. Abawi, *Plant Dis.*, 1981, **65**, 859–864.
130. J. K. Langat, J. W. Kimenju, G. K. Mutua, W. M. Muiru and W. Otieno, *Asian J. Plant Pathol.*, 2008, **7**(5), 467–472.
131. J. D. Menzies, *Phytopathology*, 1959, **49**, 648–652.
132. H. L. Jackson and E. Mtengeti, *Livestock Res. Rural Dev.*, 2005, **17**, 10.
133. J. K. Lekasi, J.C. Tanner, S.K. Kimani and P. J. C. Harris, Manure management in the Kenya highlands: practices and potential, Kenya Agricultural Research Institute, Nairobi, Kenya; ILRI Nairobi, Kenya, The Henry Doubleday Research Association (HDRA), Coventry, 1998, pp. 1–35.
134. P. L. Mafongoya, K. E. Giller and C. A. Palm, *Agroforest Syst.*, 1998, **38**, 77–97.
135. E. Ouédraogo, A. Mando and N. P Zombie, *Agric. Ecosyst. Environ.*, 2001, **84**, 259–266.
136. C. A. Palm, R. J. K. Myers and S. M. Nandwa, *Replenishing Soil Fertility in Africa,* ed. R. J. Buresh, P. A. Sanchez and F. Calhoun, SSSA Special publication Number 51, 1997, pp. 193–217.
137. M. A. Altieri, *Agric, Ecosyst Environ*, 1999, **74**, 19–31.
138. B. Clergue, Amiaud, F. Pervanchon, F. Lasserre-Joulin and S. Plantureux, *Agron. Sustainable Dev.*, 2005, **25**, 1–15.

139. B. Shiferaw, J. J. Okello and R. V. Raddy, *Environ. Dev. Sustainability.*, 2009, **11**(3), 601–619.
140. C. B. Barrett, J. Lynam, F. Place, T. Reardon and A. A. Aboud, *Natural Resource Management in African Agriculture. Undersigning and Improving Current Practices,* ed. C. Barrett, F. Place and A. Aboud, CABI Publishing, Wallinford, UK, 2002, pp. 287–296.
141. B. Shiferaw and C. Bantilan, *J. Food, Agric. Environ.*,2004, **2**(1), 328–339.
142. J. Baidu-Forson, *Agric. Econ.*, 1999, **20**, 231–239.
143. Y. Biot, P. Blakie, P. Jackson and R. Palmer-Jones, *World Bank Discussion Paper 289*, The World Bank. Washington DC, 1995.
144. W. Critchley, S. Sombatpanit and P. Madina, *Response to Land Degradation*, ed. E. Bridges, I. Hannam, I. Oldeman, L. Penning de Vries, F Scherr and S. Sombatpanit, Science Publishers Inc., Enfield, 2001.
145. S. Pandey, in *Response to Land Degradation*, ed. E. I. Bridges, I. Hannam, L. Oldeman, P. de Vries, F Scherr and S. Sombatpanit, Science Publishers Inc., Enfield, 2001.
146. R. Chambers, A. Pacey and L. Thrupp, Farmer first: Farmer Innovation and Agricultural Research, Intermediate Technology Publications, 1989, pp. 218.
147. C. Reij, C. Indigenous soil and water conservation in Africa, *Gatekeeper Series* 27, International Institute for Environment and Development, London, United Kingdom, 1991.
148. M. Robbins and T. O. Williams, presented at a STAP Workshop on Sustainable Land Management, Washington DC, 2005.
149. H. Hurni, A*gric. Ecosyst. Environ.*, 2000, **81**, 83–92.
150. D. R. Lee, *Am. J. Agric. Econom.*, 2005, ***87***, 1325–1334.
151. J. Kerr and N. Sanghi, N, Indigenous soil and water conservation in India's semi-arid tropics, International Institute for Environment and Development, 1992.
152. P. Driessen, J. Deckers, O. Spaargaren and F. Nachtergaele, *Lecture Notes on the Major Soils of the World*, World Soil Resource Reports No. 94, FAO Rome, 2001, pp. 334.
153. P. M. Wachira, S. Okoth, J. Kimenju, R. K. Mibey and J. Kiarie, *J. Trop. Subtrop. Agroecosyst.*, 2011, **13**, 43–49.
154. J. Desaeger and M. R. Rao, *Agroforest Syst.*, 2000, **47**(1), 273–288.
155. N. L. Frazer, *Mycolog. Res.*, 1999, **103**, 372–384.
156. M. P. Haware, *Options Méditerranéennes, Série Séminaires*, 1990, **9**, 63–166.
157. S. J. Muriungi, *MSc thesis*, University of Nairobi, Kenya, 1997.
158. Z. K. Punja, *Advances in Plant Pathology,* ed. G. S. Sidhu, Academic Press, London, 1988, pp. 523–534.
159. J. M. Onsando, *Tea*, 1988, **9**, 47–49.
160. J. W. Kimenju, D. M. Muiru, N. K. Karanja, M. W. Nyongesa, D. W. Miano and G. K. Mutua, *J. Trop. Microbiol.*, 2004, **3**, 14–23.
161. F. O. Ayuke, PhD thesis, Wageningen University, 2010.

Organic Matter Availability and Management in the Context of Integrated Soil Fertility Management in sub-Saharan Africa

B. VANLAUWE

International Institute of Tropical Agriculture, c/o ICIPE, PO Box 823–00621, Nairobi, Kenya
E-mail: b.vanlauwe@cgiar.org

ABSTRACT

Appropriate management of organic resources is fully embedded in the Integrated Soil Fertility Management (ISFM) framework. ISFM aims at maximising the use efficiency of external inputs through the use of improved germplasm, well-managed fertiliser and organic inputs, and adaptation of any practices to prevailing local farming conditions, including the management of non-responsive soils, or soils on which crops do not respond to fertiliser application. After a summary of the role of organic resources in tropical soil fertility management as affected by changing paradigms, the organic resource quality concept is introduced and important observations regarding the current availability and use of organic inputs in African smallholder farming systems are highlighted. The role of organic resources within ISFM is explored in the following

Issues in Environmental Science and Technology No. 35
Soils and Food Security
Edited by R.E. Hester and R.M. Harrison

Published by the Royal Society of Chemistry, www.rsc.org

ways: (i) ISFM as an entry point for producing organic resources *in situ*, the most viable mode of organic resource acquisition in African smallholder systems; (ii) the occurrence of and mechanisms underlying positive interactions between organic inputs and fertiliser, specifically focusing on the role or organic resource quality; (iii) organic resources as a solution to site-specific constraints, including high phosphorus sorption, soil acidity or soil erosion; and (iv) the potential role of organic resources in rehabilitating non-responsive soils. In a last section, the potential impact of ISFM on soil organic matter stocks and quality are addressed. In summary, although ISFM cannot be implemented without organic resources, the ISFM framework takes into account the realities of organic resource availability at the smallholder farm level.

1 Introduction

In sub-Saharan Africa (SSA) agricultural production growth is lagging behind population growth, resulting in decreasing *per capita* food production. Moreover, in contrast with other regions in the tropics, production increases for the major crops are the result of areas expansion rather than productivity increases.[1] Many reasons have been proposed that explain the lack of Green Revolution-type of productivity increases in SSA, including biophysical (*e.g.* the presence of old or shallow soils and the lack of geological soil rejuvenation), economic (*e.g.* the high price and/or unavailability of fertiliser), infrastructural (*e.g.* the absence of efficient and profitable input and output markets), social (*e.g.* the perseverance of traditional beliefs related to planting times), and political (*e.g.* the lack of government investment in agriculture), with each of these likely playing a role in the persistence of the colloquially called 'one-ton agriculture'. That said, in many areas of SSA where population densities are too high to allow for fallow-based soil fertility regeneration phases, intensification is a necessity to reduce rural poverty and hunger. In other areas with lower population densities, the rural population is often located near environmentally valuable ecosystems (*e.g.* the primary forest of the Congo basin) with arguments for intensification also applying to such areas in line with the Borlaug hypothesis that preservation of natural ecosystems can only happen through intensification of agricultural land.

Before addressing the main topic of this chapter, it is important to sketch some important characteristics of African smallholder farming systems. In the 1970s, it was concluded that African soils are as variable, if not more so, as soils in other regions.[2] Such variability strongly impacts on soil fertility and its management. At the regional scale, overall agro-ecological and soil conditions have led to diverse population and livestock densities across SSA and to a wide range of farming system. [3] Each of these systems has different crops, cropping patterns, soil management considerations, and access to inputs and commodity markets.

At the national level, smallholder agriculture is strongly influenced by governance, policy, infrastructure, and security levels. Roads also play a major role in fostering agricultural intensification through access to farm input and commodity markets. Some countries seek to control farmer associations and produce markets while others provide incentives for rural collective actions and free markets. The 'filière cotton' in Burkina Faso provides a positive example for services to members through improved access to farm technologies and product marketing.[4] Within farming communities, a wide diversity of farmer wealth classes, inequality, and production activities may be distinguished. Traditionally, local indicators of wealth have been identified that can then be used to classify farming households against a set of thresholds. In Western Kenya, farmer typologies were developed based upon production objectives of individual households, as related to their access to production factors.[5] The application of this knowledge to the process of technology adoption has been demonstrated and farmers with a larger quantity and wider diversity of resources were observed to be able to assume greater risk and venture more readily into new technologies and farm enterprises.[6]

At the individual farm level, it is important to consider the variability between the soil fertility status of individual fields, which may be as large as differences between different agro-ecological zones. This variability has obvious consequences for crop productivity, resulting in yield ranges between 900 and 2400 kg maize grain ha^{-1} for different fields within the same farm, as was documented in western Kenya.[7] These within-farm soil fertility gradients (SFG) exist most often in areas with large population densities, resulting in intensive use of land, and where amounts of farmyard manure are insufficient. SFGs are created by the position of specific fields within a 'soil-scape', by the selective allocation of available nutrient inputs to specific crops and fields, and by improved management (*e.g.* time of planting, weeding, *etc.*) of plots with higher fertility.[7] An important proportion of soils in such regions are often described as 'non-responsive', signifying that crops grown on such soils do not respond to regular NPK fertiliser application.[8] In cases where such soils represent a substantial proportion of the total agricultural land area, options to rehabilitate these are crucial towards agricultural intensification.

The objectives of this chapter are: (i) to sketch the historical logic that led to the conceptualisation of ISFM; (ii) to highlight the role of organic resource management within the ISFM framework; and (iii) to address potential linkages between ISFM practices and the soil organic matter pool. These objectives are addressed within the context of areas where intensification is a must and where ISFM is a potential way forward to reach intensification. Areas with relatively long fallow periods are thus not the focus of this paper.

2 Organic Matter in Relation to Paradigm Shifts in Tropical Soil Fertility Management

Over the years, several paradigms have been developed and evaluated to address the intensification question. During the 1960s and 1970s, an external input paradigm was driving the research and development agenda (see Table 1). The

Table 1 Changes in tropical soil fertility management paradigms over the past 5 decades and the specific role of fertiliser and organic inputs in the context of these paradigms.

Period	*Paradigm*	*Role of fertiliser*	*Role of organic inputs*	*Experiences*
1960s and 1970s	External Input Paradigm 'First Paradigm'	Use of fertiliser alone will improve and sustain yields.	Organic resources play a minimal role.	Limited success due to shortfalls in infrastructure, policy, farming systems, *etc.*
1980s	Organic Input Paradigm	Fertiliser plays a minimal role.	Organic resources are the main source of nutrients.	Limited adoption; organic matter production requires excessive land and labour.
1990s	Sanchez' 'Second Paradigm'	Fertiliser use is essential to alleviate the main nutrient constraints.	Organic resources are the entry point; these serve other functions besides nutrient release.	Difficulties to access organic resources hampered adoption (*e.g.* improved fallows).
2000s	Integrated Soil Fertility Management	Fertiliser is a major entry point to increase yields and supply needed organic resources.	Access to organic resources has various social and economic dimensions.	On-going; several success stories.

appropriate use of external inputs, be it fertilisers, lime, or irrigation water, was believed to be able to alleviate any constraint to crop production. Following this paradigm, together with the use of improved cereal germplasm, the 'Green Revolution' boosted agricultural production in Asia and Latin America in ways not seen before. Organic resources were considered less essential.

Because the 'Green Revolution' by-passed sub-Saharan Africa (massive applications of fertilisers and pesticides resulted in environmental degradation in Asia and Latin-America, and fertiliser subsidies were abolished in SSA, imposed by structural adjustment programs), a renewed interest in organic resources could be observed in the early 1980s. The balance shifted from mineral inputs only to low mineral input sustainable agriculture (LISA), where organic resources were believed to enable sustainable agricultural production. Because LISA systems did not emphasise the need for fertiliser, organic resources were merely considered as short-term sources of nutrients, especially N. After a number of years of investment in research activities evaluating the potential of LISA technologies, such as alley cropping or live-mulch systems, several constraints were identified both at the technical (*e.g.* lack of sufficient organic resources) and the socio-economic level (*e.g.* labour-intensive technologies).

In this context, the Second Paradigm for tropical soil fertility research was formulated as: '*Rely more on biological processes by adapting germplasm to adverse soil conditions, enhancing soil biological activity and optimising nutrient cycling to minimise external inputs and maximise the efficiency of their use*' (see Table 1).[9] This paradigm did recognise the need for both mineral and organic inputs to sustain crop production, and emphasised the need for all inputs to be used efficiently. The need for both organic and mineral inputs was advocated because (i) both resources fulfil different functions to maintain plant growth; (ii) under most small-scale farming conditions, neither of them is available or affordable in sufficient quantities to be applied alone; and (iii) several hypotheses could be formulated leading to added benefits when applying both inputs in combination. The second paradigm also highlighted the need for improved germplasm; as in earlier days, more emphasis was put on the nutrient supply side without worrying too much about the demand for these nutrients. Obviously, optimal synchrony or use efficiency requires both supply and demand to function optimally.

The need for sustainable intensification of agriculture in SSA and the recognition that this cannot happen without fertiliser has recently gained strong support, in part because of the growing recognition that farm productivity is a major entry point to break the vicious cycle underlying rural poverty. Recent landmark events include the African Heads of State Fertiliser Summit held in Abuja, Nigeria,[10] and the launching of the Alliance for a Green Revolution in Africa (AGRA). Kofi Annan, the chairman of the board of AGRA, has repeatedly stressed that the African Green Revolution should be uniquely African by recognising the continent's great diversity of landscapes, soils, climates, cultures and economic status, while also learning lessons from earlier Green Revolutions in Latin America and Asia.[11] Since fertiliser is an expensive commodity, AGRA has adapted Integrated Soil Fertility Management (ISFM) as a framework for boosting crop productivity through reliance upon soil fertility management technologies, with emphasis on increased availability and use of mineral fertiliser. ISFM is defined as '*A set of soil fertility management practices that necessarily include the use of fertiliser, organic inputs, and improved germplasm combined with the knowledge on how to adapt these practices to local conditions, aiming at maximising agronomic use efficiency (AE) of the applied nutrients and improving crop productivity. All inputs need to be managed following sound agronomic principles*'.[8] Conceptually, ISFM can be sketched as shown in Figure 1 which explicitly addressed the need for non-responsive soils to be rehabilitated, *e.g.* through the use of organic resources, before an increase in fertilier AE can be expected (Paths B to C in Figure 1).

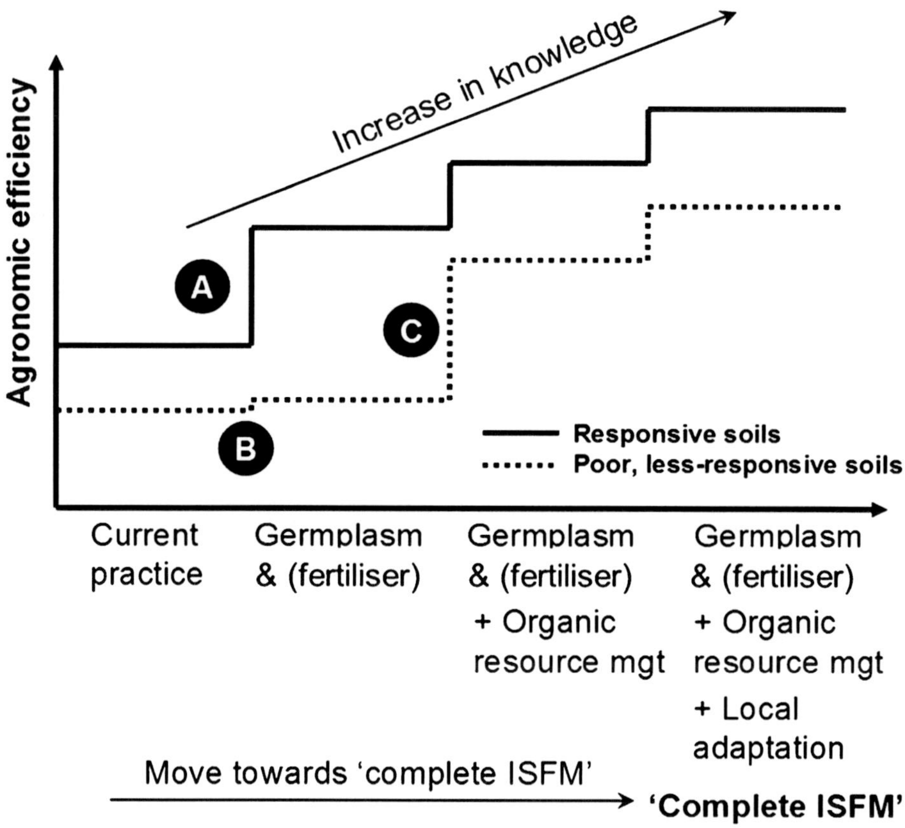

Figure 1 Conceptual relationship between the agronomic efficiency (AE) of fertiliers and organic resource and the implementation of various components of ISFM, culminating in complete ISFM towards the right side of the graph. Soils that are responsive to NPK-based fertiliser and those that are poor and less-responsive are distinguished. The 'current practice' step assumes the use of the current average fertiliser application rate in SSA of 8 kg fertiliser nutrients ha^{-1}. The meaning of the various steps is explained in detail in the text. At constant fertiliser application rates, yield is linearly related to AE.[8]

ISFM is derived from Sanchez's earlier 'Second Paradigm' but uses fertiliser as the entry point for improving crop productivity. It asserts that substantial and necessary organic resources may be derived as by-products of food crops and livestock enterprise. ISFM also recognises the importance of an enabling environment that permits farmer investment in soil fertility management, and the critical importance of farm input suppliers and fair produce markets, favourable policies, and properly functioning institutions, particularly agricultural extension.

3 Availability and Quality of Organic Resources in African Farming Systems

Before addressing the main theme of this chapter, it is important to stress some important notions related to organic matter quality and availability in African farming systems, since these determine some of the boundary conditions within which ISFM necessarily resides.

3.1 The Organic Resource Quality Concept

Although organic inputs had not been new to tropical agriculture, the first seminal synthesis on organic matter management and decomposition was written only in 1979.[12] Between 1984 and 1986, a set of hypotheses was formulated based on two broad themes, 'synchrony' and 'soil organic matter (SOM)', building on the concepts and principles formulated in 1979. Under the first theme, especially the O(Organisms)-P(Physical environment)-Q(Quality) framework for OM decomposition and nutrient release, formulated earlier, was worked out and translated into hypotheses driving management options to improve nutrient acquisition and crop growth. Under the second theme, the role of OM in the formation of functional SOM fractions was stressed. During the 1990s, the formulation of the research hypotheses related to residue quality and N release led to a vast number of projects aiming at validation of these hypotheses, commonly resulting in meaningful relationships between N release dynamics and the organic resource quality, expressed as various combinations of its N, soluble polyphenol and lignin contents.

Two major events further accentuated the relevance of organic resource quality in tropical soil fertility management. Firstly, a workshop was held in 1995 with its theme 'Plant litter quality and decomposition' resulting in a book summarising the state of the art of the topic.[13] Secondly, the Tropical Soil Biology and Fertility Institute (TSBF) in collaboration with its national partners and Wye College developed the Organic Resource Database (ORD) and related Decision Support System (DSS) for OM management.[14] The Organic Resource Database contains information on organic resource quality parameters, including macronutrient, lignin and polyphenol contents of fresh leaves, litter, stems and/or roots from almost 300 species found in tropical agro-ecosystems. Careful analysis of the information contained in the ORD led to the development of the DSS, which makes practical recommendations for appropriate use of organic materials based on their N, polyphenol and lignin contents, resulting in four categories of materials (Classes I to IV). The DSS recognises the need for certain organic resource to be applied together with mineral inputs, consistent with the Second Paradigm.

3.2 Availability and Production of Organic Resources in African Farming Systems

In areas with absence of relatively long fallow periods, available organic resources usually consist of (a combination of) crop residues, animal manure, composted household waste, and prunings from trees or shrubs within plots or along their borders. Due to the low current productivity of farming systems, any of the above is usually available in limited quantities. Manure availability is largely determined by livestock densities, grazing and feeding systems, and storage measures.[15] In Western Kenya, for instance, it was observed that the use of organic inputs decreased strongly with the distance from the homestead and/or grazing sites, and differed between the crops.[7] Vegetable crops grown in home gardens received most of the organic resources, followed by the cash and grain crops grown in the close and mid-distance fields. Virtually no organic resources were applied to the remote fields, due to the extra effort required to transport coarse materials to distant parts of the farm. Moreover, in many areas, crop residues have competing uses and are often removed from the land to serve as livestock feed or fencing material or to be sold to other farmers. In the West-African Sahel, herdsmen are often allowed to graze on crop residues through specific arrangements between sedentary farmers and pastoralists.[16] Organic resource availability is also different for different farm typologies with poorer farmers having less access to organic inputs and often still selling part of whatever resources they have to better-off farming families.[5]

The lack of organic inputs within farms and the need to use such inputs for restoring soil fertility has been recognised for a long time and several attempts have been made to increase the availability of organic resources within smallholder farms. Several windows in time and space can be identified to produce organic resources (see Table 2) and all of these have been the focus of specific measures, often referred to as ‘improved systems’, over the past three decades. Simultaneous *in situ* production of organic matter has been most widely studied in agroforestry systems. In the West-African savannah, for instance, hedgerow trees in alley cropping systems were observed to produce 940–6027 kg dry matter ha^{-1} $season^{-1}$.[17] After many years of adaptation and promotion of agroforestry systems, one could conclude: (i) that several agroforestry systems are technically sound, but only for specific soil fertility, rainfall and crop conditions, and (ii) that several constraints to adoption of agroforestry systems exist, related to land tenure, the time for such systems to deliver the expected goods and services, the need for extra labour, and the lack of appropriate seed systems, amongst others.[18,19] Long-duration herbaceous

Table 2 Place and time of production of organic matter (fallow species) relative to crop growth and the respective advantages/disadvantages of the mentioned organic matter production systems with respect to soil fertility management and crop growth. 'Same place' and 'same time' mean 'in the same place as the crop' and 'during crop growth'.

Place and time of organic matter production - example of farming system	*Advantages*	*Disadvantages*
Same place, same time	- 'Safety-net' hypothesis (complementary rooting depths)	- Potential competition between crop and fallow species
- Alley cropping	- Possible direct transfer from N_2 fixed by legume species	- Reduction of available crop land
Same place, different time	- 'Rotation' effects (N transfer, improvement of soil P status, *etc.*)	- Land out of crop production for a certain period
- Crop residues	- *In situ* recycling of less mobile nutrients	- Decomposition of organic matter may start before crop growth (potential losses of mobile nutrients, *e.g.* N, K)
- Legume-cereal rotation	- No competition between fallow species and crops	- Extra labour needed to move organic matter (manure)
- Manure, derived from livestock fed from residues collected from same field		
Different place	- Utilisation of land/ nutrients otherwise not used	- Extra labour needed to move organic matter
- Cut-and-carry systems	- No competition between fallow species and crops	- No recycling of nutrients on crop land
- Household waste		- Need for access to extra land
- Animal manure, not originating from same field		- Manure and household waste often have low quality

legumes, such as *Mucuna pruriens* (L.) var *utilis* (Wright) Burck, were observed to produce a similar range of biomass values (1800–8700 kg dry matter ha^{-1} $season^{-1}$) under similar agro-ecological conditions. [17] Major observations related to such systems are: (i) herbaceous fallow cover cropping systems are a technically sound system under most agro-ecological conditions, provided the soil fertility status is not degraded below specific thresholds; (ii) as with agroforestry systems, herbaceous legumes require immediate investments in terms of land and labour, while their benefits are only seen in a following season; and (iii) specific niches exist for herbaceous legume-based technologies, for instance for the control of *Imperata cylindrica* L. weeds or the supply of livestock feed, but their adoption has been very limited in scale notwithstanding many years of investment and promotion. Cut-and-carry systems, including *Tithonia diversifolia* (Hemsl.) A. Gray biomass transfer systems, have the advantage that competition between the fallow species and the crop is excluded, but the main disadvantage that additional labour is required to move the organic inputs from their source to the cropped field.[20]

Grain legumes, such as cowpea or soybean, usually produce less biomass and a significant amount of nutrient is removed with the harvested products, but these are traditionally part of most cropping systems. Although improved varieties of these grain legumes have a great potential to be adopted by farmers, the earlier-developed germplasm contributed little to improving the soil fertility status because their biomass accumulation was low and/or their N harvest index was high and larger than the proportion of N fixed from the atmosphere, leading to net negative contributions to the soil N balance. However, over the past decade, grain legume breeding programmes have been shifting from maximising grain yield alone to maximising grain yield and fodder production – the so-called 'dual purpose' legumes. Such varieties usually fixed more N than was exported with the grains and left a significant amount of N in the soil to be potentially taken up by a following cereal. Such a variety is, for example, TGX-1448-2E that produced between 470 and 2080 kg grains ha^{-1} (average of 1290 $\pm$500 kg ha^{-1}) and between 1000 and 5340 kg biomass ha^{-1} at peak biomass (average of 2510 $\pm$1050 kg ha^{-1}), and fixed between 78% and 92% of its N (average of 84 $\pm$4%) when grown on 27 farmers' fields in Northern Nigeria. Not surprisingly, maize growing after these improved soybean varieties had significantly higher grain yield (1.2–2.3-fold increase) compared to a maize control.[21] The most promising options in terms of farmers' interest and adoption potential are based on the integration of such multi-purpose grain legumes into existing farming systems either through system adaptation, *e.g.* by adapting plant spacing to allow for higher legume densities, or diversification, *e.g.* through inclusion of alternative legumes.[22]

4 Organic Matter Production and Use in the Context of ISFM

Integrated Soil Fertility Management (ISFM) aims at intensifying agricultural production through integration of essential and promising components, thereby acknowledging the constraints smallholder farmers face in terms of land, labour, and minimising modifications to existing farming systems. Organic inputs are closely linked to ISFM in a several ways: (i) ISFM has the potential to increase the availability of organic resources at farm level; (ii) organic resources have the potential to enhance fertiliser use efficiency and rehabilitate non-responsive soils; and (iii) production systems operating at higher levels of organic resource throughput can retain more soil organic matter, with potentially positive impacts on fertiliser use efficiency and system productivity.

4.1 Organic Matter Production through ISFM

The first components of the ISFM strategy consist of improved germplasm in combination with target and well-managed fertiliser application. Applying fertiliser to germplasm that is unresponsive, not adapted to a specific environment, or that is affected by pests and diseases, will result in low demand for nutrients, low productivity, and thus low AE values. Higher crop yields are related with a larger amount of crop residues that can then be either fed to livestock, with manure potentially returned to the field, and/or recycled *in situ* towards improving fertiliser AE and enhancing the soil organic matter pool. In South Kivu, DR Congo, for example, improved, open-pollinated maize varieties yielded more than local varieties without fertiliser and had a higher response to fertiliser application, resulting in higher AE values (see Figure 2). In the same region, yields of maize following soybean or climbing beans (*Phaseolus vulgaris* L.) were 27–57% higher than that of maize following maize (see Figure 3). Rotational benefits were also greater when improved, dual-purpose legume varieties with a low harvest index were grown. These legumes gave similar grain yields to local varieties (not shown), but grain yields of following maize crops were 20–34% higher than those of maize following local legume varieties. These yield improvements were related to greater biological nitrogen (N) fixation in the improved legumes, which derived a greater proportion of N from the atmosphere (due to their longer growing period relative to local varieties) and gave a higher biomass yield.

4.2 Interactions between Organic Matter and Fertiliser

The second component of ISFM advocates the combined application of organic inputs and fertiliser. Organic inputs contain nutrients that are released at a rate determined in part by their chemical characteristics or organic resource quality. However, organic inputs applied at realistic levels seldom release sufficient nutrients for optimum crop yield. Combining organic and mineral inputs has been advocated as a sound management principle for smallholder farming in the tropics because neither of the two inputs is usually available in sufficient quantities and because both inputs are needed in the long-term to sustain soil fertility and crop production. Furthermore, positive interactions between fertiliser and organic inputs have frequently been observed resulting in extra grain yield[23] or reduced N losses.[24] In the earlier example (Figure 3) for instance, combining crop rotation and fertiliser application resulted in yield increases up to 120% relative to the unfertilised maize-maize rotation, and a mean fertiliser value : cost ratio of 2.7:1.

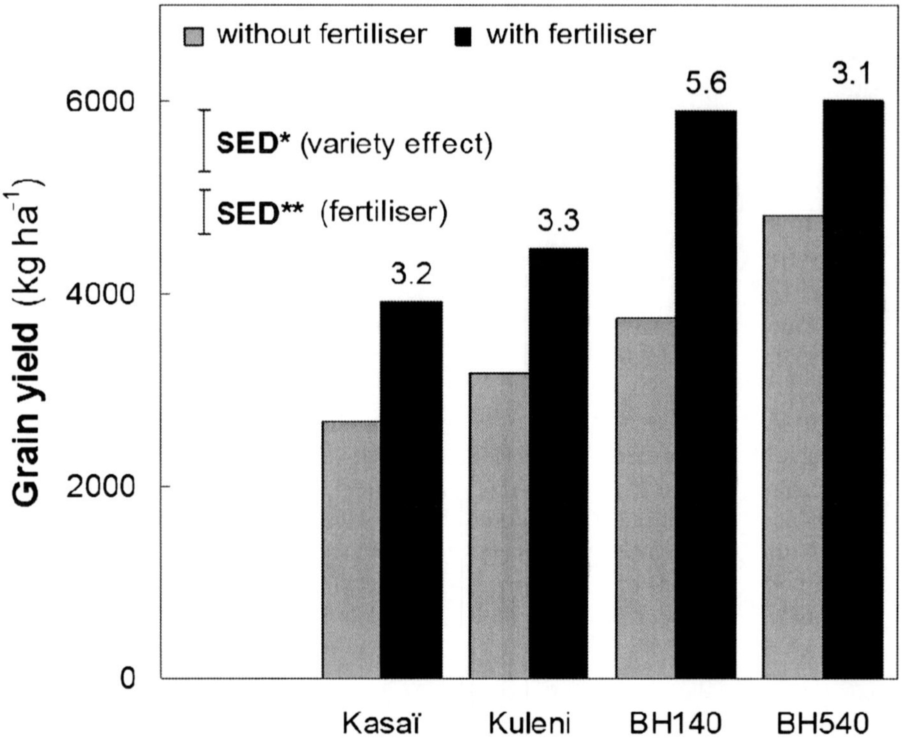

Figure 2 Grain yield of two local maize varieties (Kasaï and Kuleni) and two improved, open-pollinating maize varieties (BH140 and BH540) as affected by application of 13 kg phosphorus, 60 kg nitrogen, and 25 kg potassium per hectare across four sites in South Kivu, Democratic Republic of the Congo. Values above yields with fertiliser represent the value : cost ratios [in USD per USD], assuming fertiliser unit prices of 1.7 and 0.9 \$ kg^{-1} for NPK and urea, and 1 \$ kg^{-1} for maize grain. SED = standard error of difference; * = significant at $P < 0.05$; ** = significant at $P < 0.01$. There was no significant fertiliser × variety interaction.[22]

In 2001, a 'Direct' and an 'Indirect Hypothesis' were formulated, underlying this phenomenon.[17] For N fertiliser, the 'Direct Hypothesis' may be formulated as: '*Temporary immobilisation of applied fertiliser N may improve the synchrony between the supply of and demand for N and reduce losses to the environment*'. The 'Indirect Hypothesis' may be formulated for a certain plant nutrient X supplied as fertiliser as: '*Any organic matter-related improvement in soil conditions affecting plant growth (except the nutrient X) may lead to better plant growth and consequently enhanced efficiency of the applied nutrient X*'. Translated to N, these hypotheses are based on direct and indirect interactions between organic inputs and N fertiliser. The former are governed by microbial processes and influence the supply of plant-available N directly, leading to improved synchrony between the supply of and demand for N. The latter indirect interactions, which can occur simultaneously with the direct

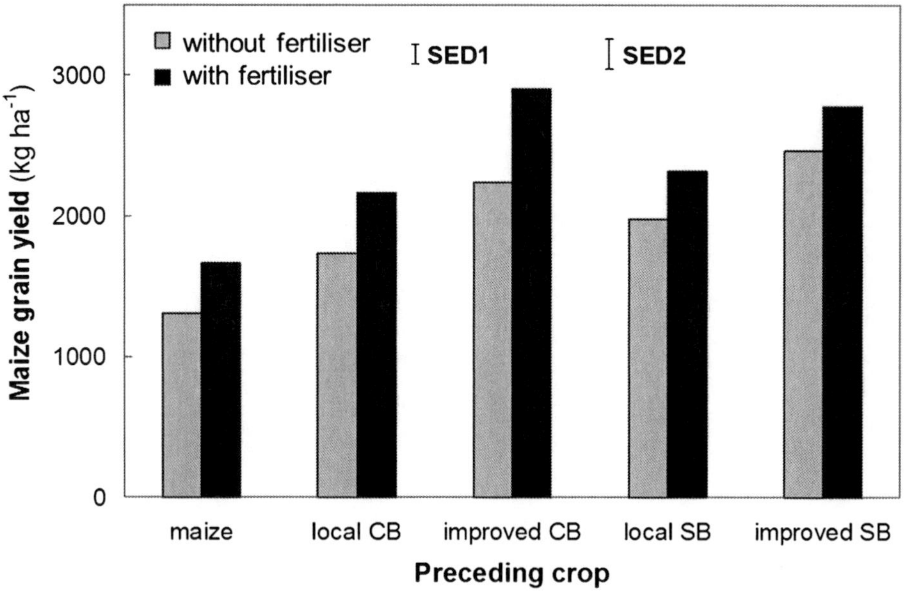

Figure 3 Maize grain yield as affected by application of compound fertiliser (NPK, 17 : 17 : 17) at 100 kg ha^{-1} and the crop grown in the preceding season (maize, climbing beans [CB] or soybean [SB]) in South Kivu, DR Congo. SED = standard error of difference; * = significant at $P < 0.05$; ** = significant at $P < 0.01$. There was no significant fertiliser × preceding crop interaction. CB: climbing bean (*Phaseolus vulgaris* L.); SB: soybean (*Glycine max* L.).[22]

interactions, originate in organic input-driven alleviation of other growth-limiting factors besides N. Consequently they rather influence the demand for plant-available N and lead to higher N uptake. They do not necessarily, however, improve synchrony as the timing of such improved demand may not have changed drastically. Through a meta-analysis, fertiliser N-AE values were observed to be significantly higher for the treatments where fertiliser was combined with manure or compost [38 kg (kg N)$^{-1}$], but all other organic resources did not significantly affect N-AE values compared to the sole fertiliser treatment [25 kg (kg N)$^{-1}$].[25] However, when performing the statistical analysis on the data with maximum organic N application rates of 30 or 60 kg N ha^{-1}, organic inputs belonging to Class II and manure/compost had significantly higher N-AE values than the sole fertiliser treatment or the Classes I and III/IV organic inputs (Figure 4). At higher organic N application rates, only the treatment with manure/compost gave significantly higher N-AE values than the sole fertiliser treatment (Figure 4). This confirms that higher quality organic inputs are mainly a source of nutrients while Class II or manure/compost alleviated other constraints to plant growth besides N, thus improving the uptake of fertiliser N by maize.

An important issue related to organic inputs and ISFM is concerned with the question whether fertiliser application can generate the required crop

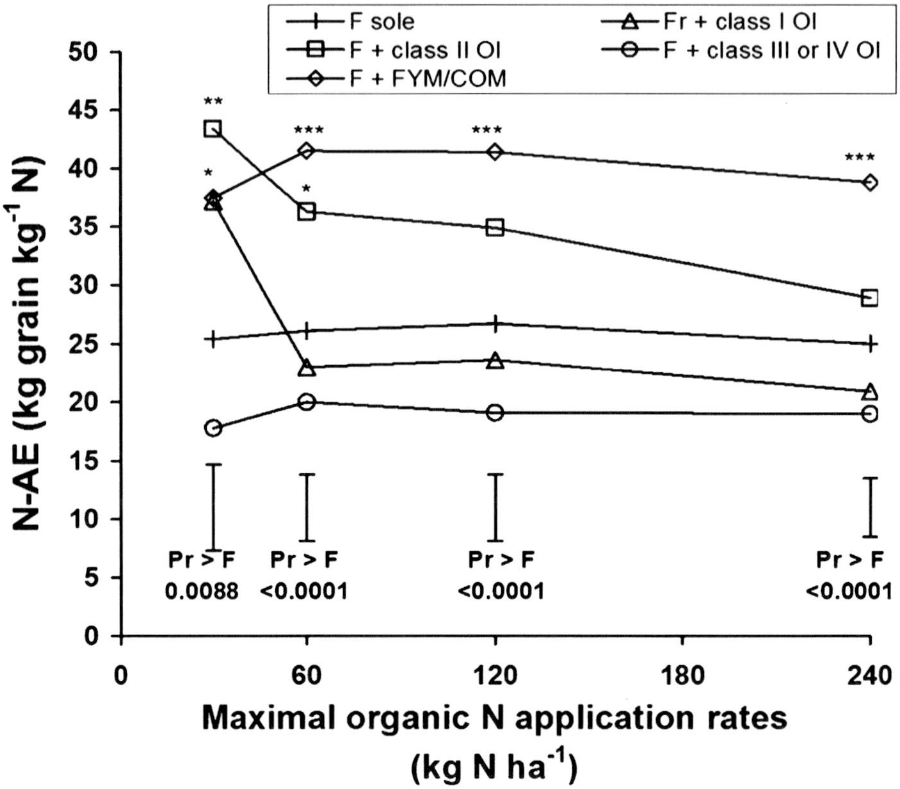

Figure 4 Agronomic efficiency of fertiliser N (N-AE) as affected by combination with different classes of organic inputs (Classes I, II, III+IV, and manure+compost) for organic N application rates ≤30 kg N ha^{-1} (125 observations), ≤60 kg N ha^{-1} (238 observations), ≤120 kg N ha^{-1} (305 observations), or ≤240 kg N ha^{-1} (352 observations). Error bars are average Standard Errors of the Difference. The symbols *, **, and *** indicate a significant difference with the sole fertiliser treatment at the 0.1, 1, and 5% level. In the legend, F, OI, FYM, and COM refer to fertiliser, organic inputs, manure, and compost, respectively.[25]

residues that are needed to optimise the AE of fertiliser for a specific situation. Data obtained in Niger support this proposition.[16] Where fertiliser was applied to millet, sufficient residue was produced to meet both farm household demands for feed and food as well as the management needs of the soil in terms of organic inputs and surface protection of the soil from wind erosion.

4.3 Organic Matter and Local Adaptation

The third component of ISFM addresses local constraints to improved fertiliser AE. At this level, organic inputs have been shown to reduce P sorption, at least in over one growing season,[26] and alleviate soil acidity-related constraints when concentrated in the planting hole. For instance, the placement of high quality manure in the planting hole was not only found to increase climbing bean yields on Ferralsols in East DR Congo from less than 500 to over 1500 kg ha^{-1}, but also increase the response to NPK fertiliser application (see Figure 5). Both effects were much less pronounced when the manure was broadcast over the whole plot surface (Figure 5).

Organic inputs applied as mulch gain prominence nowadays in the context of the Conservation Agriculture debate.[27,28] Conservation Agriculture advocates the use of three principles as a pathway towards sustainable agriculture: (i) minimal or reduced tillage, (ii) retention of surface mulch to keep the soil covered, and (iii) crop rotation and diversification. While it is not the intention of this chapter to join this debate, organic resources applied on the surface have been shown to improve soil moisture conditions, resulting in higher and more stable yields.[23,29] There is less consensus, however, related to interactions between surface mulch and fertiliser use efficiency since surface mulch in the absence of tillage results in higher soil moisture contents and more continuous soil pores, potentially resulting in larger fertilier N losses due to enhanced leaching. In lysimeters installed in Nigeria, for instance, it was

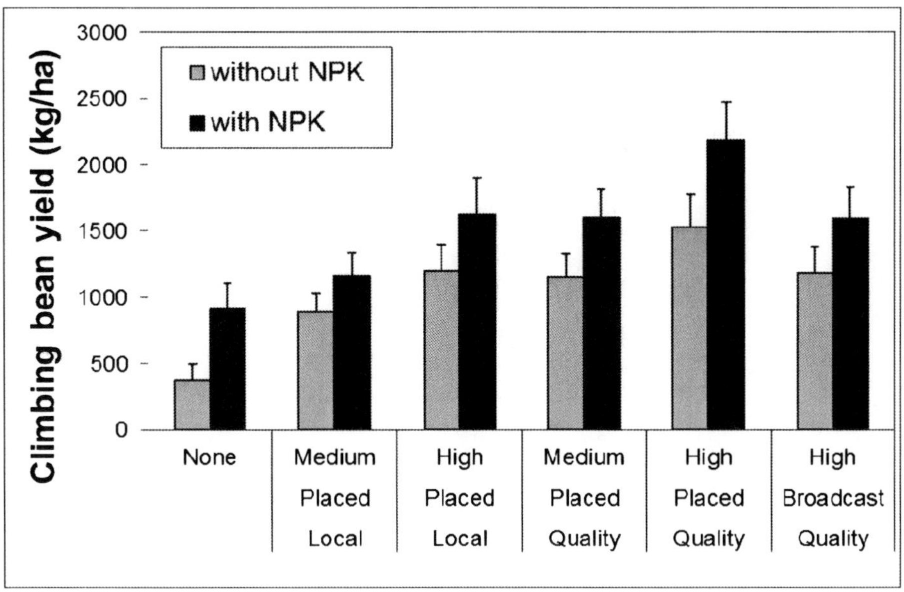

Figure 5 Climbing bean (variety AND10) yield in East DR Congo as affected by fertiliser application (with and without NPK), manure application rates (medium: 2.5 tonne dry matter ha^{-1}; high: 7.5 tonne dry matter ha^{-1}), manure placement (placed in the planting hole and broadcast), and manure quality (local and 'quality' manure). Error bars are Standard Errors of the Mean.

observed that losses of ^{15}N-labelled fertiliser were higher when maize stover was surface-applied compared with its incorporation in the topsoil.[30]

Soil erosion is a major problem in most of the Central and East African highlands. Several methods, both biological (*e.g.* live hedges along the contour lines) or physical (*e.g.* physical terrace creation) or a combination of both, have been used to reduce the removal of fertile topsoil in hilly landscapes with variable success, mostly related to the substantial labour (especially for the establishment of physical structures) requirements and the relatively long period of time before benefits can be observed (especially for biological approaches). Although within such systems, surface application of organic inputs as mulch could reduce soil erosion, evidence suggests that the largest impact on soil erosion is related to the presence of erosion control structures more than applied surface mulch. For instance, in Central Kenya, more pronounced differences were observed in soil loss between different cropping seasons than between tillage treatments (see Figure 6). Across anti-erosion barriers, cumulative soil loss was smaller with minimum than regular tillage for 2 of the 4 seasons. In the last season of the study, the short rainy season of 2008 (SR 08), soil loss was independent of tillage practice without anti-erosion barriers. With vegetative barriers, soil loss was greater for leucaena (larger for

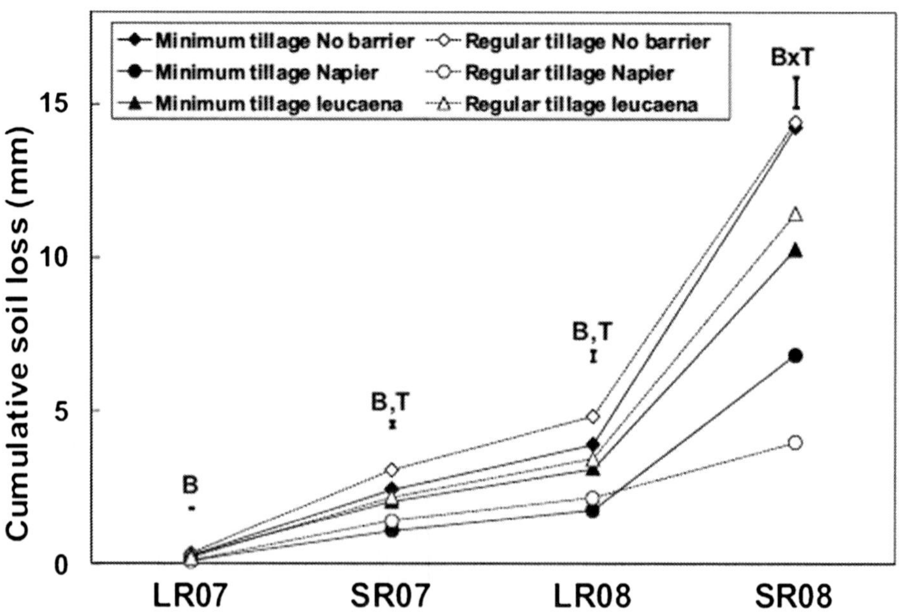

Figure 6 Cumulative soil loss for four consecutive seasons (long rainy season 2007 – short rainy season of 2008) as affected by tillage and anti-erosion barriers. A soil loss of 1 mm is equivalent to 1 kg m^{-2} or 10 Mg ha^{-1} as bulk density is not affected by treatments. Error bars represent SED for tillage (T) × anti-erosion barrier interaction.[34]

minimum than regular tillage) than with Napier barriers (larger for regular than minimum tillage) (Figure 6).

4.4 Organic Matter and Rehabilitation of Non-Responsive Soils

The nature of non-responsiveness can consist of chemical (*e.g.* soil acidity-related constraints), physical (*e.g.* the occurrence of hard pans at shallow depth), or biological (*e.g.* presence of a large *Striga hermonthica* (Delile) Benth. seed bank in maize-based systems) constraints, or a combination of all these. In areas where a large proportion of the soils occur in a non-responsive status and where soils are not irreversibly degraded (*e.g.* in cases where all topsoil has been lost due to erosion), rehabilitation of such soils is a necessity for sustainable agricultural intensification. Obviously, organic resources are not a cure to all the above constraints but, under certain circumstances, application of organic inputs, often at high seasonal or yearly rates, is an option to rehabilitate non-responsive soils and induce substantial increases in fertiliser AE (Path C in Figure 1).

In Zimbabwe, for instance, applying farmyard manure for 3 years to sandy soils at relatively high rates enabled a clear response to fertiliser where such response was not visible before rehabilitation (see Figure 7). Application of single super phosphate (SSP) fertiliser or manure with 100 kg N ha^{-1} on the sandy homefield increased maize yields in the first season, but optimum responses were attained at P application rates of 10 kg ha^{-1}. The response of maize to 100 kg N ha^{-1} at the different rates of P in the first season was very poor and not significant on the sandy outfield, with maximum yields less than 1 tonne ha^{-1}, irrespective of the source of P. The lack of response to nutrients added and dolomitic lime observed on the sandy outfield, which was acidic and low in all nutrients, was unexpected and indicates that this field was deficient in other nutrients besides N, phosphorus (P), calcium (Ca), and magnesium (Mg). Yields on the sandy outfield were marginally improved in the second season of manure application. Only in the third season did manure significantly increase yields, against decreases in yields for the sole N treatment. The yields remained markedly small on the sandy outfield with all SSP treatments (Figure 7).

In a set of medium-to-long term agroforestry trials in West Africa, topsoil Ca content, effective cation exchange capacity and pH were substantially higher under *Senna siamea* than under *Leucaena leucocephala*, *Gliricidia sepium*, or the no-tree control plots in sites with a clay accumulation soil horizon (Bt horizon) rich in exchangeable Ca.[31] This was shown to be largely related to the recovery of Ca from the subsoil under *Senna* trees. At one of the sites in Benin Republic, after 6 years of continuous cropping, topsoil degradation related to high soil acidity and low base saturation levels, resulted in zero maize yields even in presence of fertiliser.[32] Integration of hedgerow trees, especially *Senna siamea*, resulted in maize yields approaching 2 tonne ha^{-1}, indicating that hedgerow prunings can counteract soil acidity-related constraints and restore soil responsiveness.

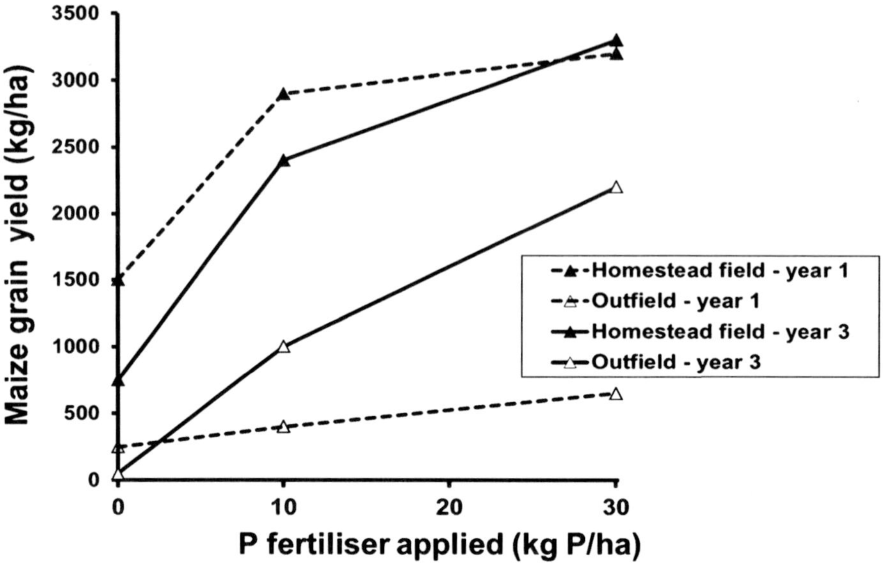

Figure 7 Maize grain yields from the outfield of the farm on the sandy soil amended with 100 kg N ha^{-1} and manure or single super phosphate in the first and third season. Bars show SEDs.[35]

5 Soil Organic Matter Status and Quality as Affected by ISFM

A last important issue of this chapter deals with the potential of ISFM to enhance the soil organic matter (SOM) pool and the various services this pool provides, including cation exchange capacity, rainfall infiltration, and soil structure. Before addressing this issue, it is important to note that there is a growing body of evidence that the longer-term impact of organic resource quality (see section 3.1.) is not affecting the quantity or the quality of the soil organic matter pool. In a field trial in Central Kenya, for instance, it was demonstrated that litter quality did not influence longer-term SOC stabilisation (see Figure 8). Organic input addition was the only factor that influenced SOM concentrations, although the inputs used varied widely in their N, lignin and polyphenol contents, and covered the range of litter qualities commonly available for field amendment in tropical agro-ecosystems. Regardless, these differences in litter composition did not translate into differences in SOC stabilisation. In other words, for increasing the SOM pool, application rates are far more important than the quality of the organic inputs applied. An exception to this trend appears to be manure or compost which tends to have a larger proportion of its C retained in the SOM pool, probably because these resources have passed through a decomposition process in the rumen of livestock and/or after storage before field application.

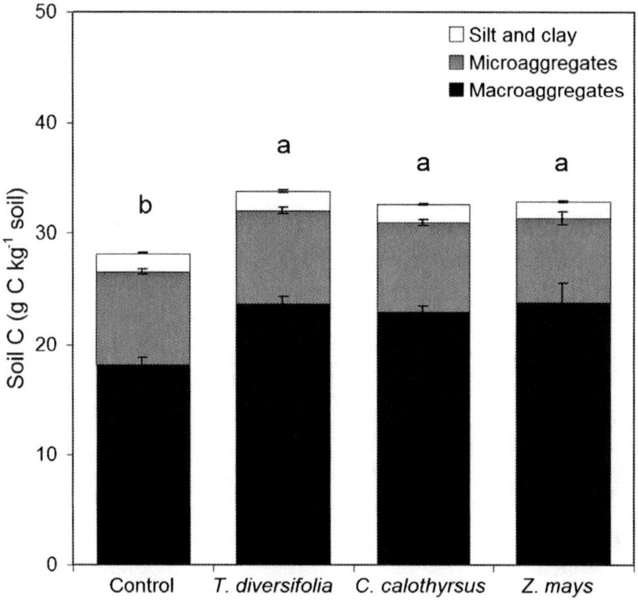

Figure 8 Soil C contents of three aggregate size fractions (macroaggregates (>250 μm), microaggregates (53–250 μm), and silt and clay (<53 μm)) after 3 years of 4 Mg litter-C ha^{-1} yr^{-1} input (no input, *Tithonia diversifolia*, *Calliandra calothyrsus*, and *Zea mays*) in a maize cropping system in Central Kenya. Error bars represent the standard error of the mean of each aggregate size fraction. Significant differences in total soil carbon between treatments are denoted by different letters.[36]

The overall aim of ISFM is maximal AE of the inputs used and thus maximal crop yields and crop residue production for a given amount of inputs along the linear part of a standard fertiliser response curve. Within the ISFM framework, crop residues are proposed to be recycled *in situ* or returned to the field in the form of manure after these have been fed to livestock. In principle, higher crop productivity should generate a larger amount of crop residues and thus have a positive effect on SOM stocks. The often-reported higher SOM contents in plots near the homestead compared with those further away are related to higher crop productivity on those plots combined with a preferential allocation of organic waste to these plots.[33]

An interesting research issue, relevant for ISFM and related to the local adaptation component of ISFM, is whether fertiliser N-AE is higher on soils with a high soil fertility status, such as the homestead fields, compared with soils with lower soil fertility status. In plots where any of the above constraints limit crop growth, a higher SOM content may enhance the demand by the crop for N and consequently increase the fertiliser N use efficiency. On the other hand, SOM also release available N that may be better synchronised with the demand for N by the plant than fertiliser N and consequently a larger SOM pool may result in lower use efficiencies of the applied fertiliser N. A

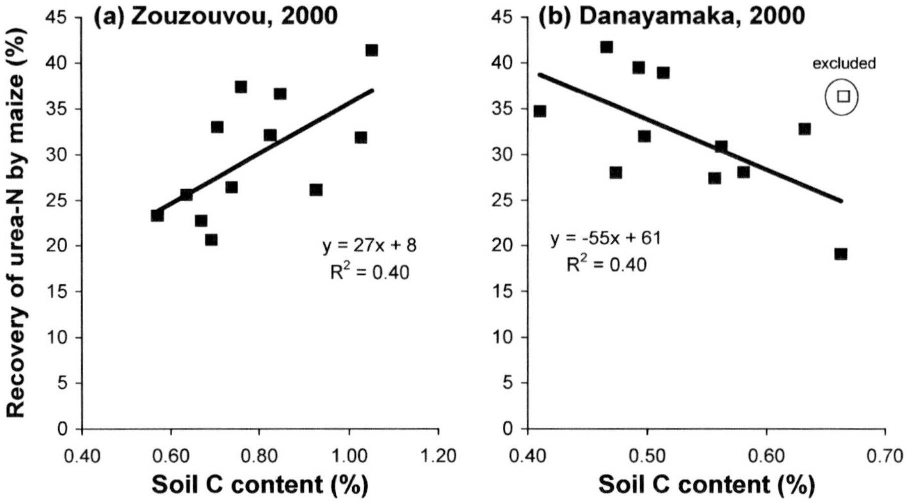

Figure 9 Observed relationships between recovery of ^{15}N-labelled urea N in the maize-shoot biomass and the soil organic-C content for 12 farmers' fields in Zouzouvou (Southern Benin) and Danayamaka (Northern Nigeria). Urea was split-applied (one third at planting, two thirds at knee height) at 90 kg N ha^{-1} in Zouzouvou and 120 kg N ha^{-1} in Danayamaka. One observation was excluded from the regression analysis for the Danayamaka data.[37]

preliminary investigation using ^{15}N-labelled urea under on-farm conditions in Southern Benin and Northern Nigeria revealed contrasting trends between the two sites (see Figure 9). Total recoveries of urea-N covered the same range in both areas (between 20 and 45%), but in Benin recovery of applied urea-N was positively related to soil organic C-content while in Nigeria a negative relationship was observed. Although the exact reasons underlying these different trends are not clear, the major function of the SOM pool in Benin is probably mainly to alleviate one or more specific constraints to crop growth besides N, while in Nigeria SOM mainly supplies N to the growing crop.

6 Conclusions

Integrated Soil Fertility Management and organic matter management are both conceptually and practically fully intertwined. Due to the low productivity of African smallholder farming systems, it is important first of all to generate sufficient organic resources within the farm. ISFM has the potential to generate these resources through the use of improved germplasm and well-managed fertiliser. Once the organic resources are available, they can be used in combination with fertiliser to enhance the use efficiency of the latter or to alleviate specific constraints to higher fertiliser use efficiency, including high P sorption, soil acidity, or soil erosion on hillsides. In the longer term, ISFM is also expected to have a positive effect on the soil organic matter pool, especially since the quantity of the organic resources applied to the soil appears to be more important than their quality.

Though technically sound, ISFM will only generate impact at the smallholder farm level if combined with initiatives that create an enabling environment for its uptake, including profitable access to input and output markets, knowledge transfer using various means, and value chain partnerships.

Acknowledgments

The authors wish to acknowledge the Belgian Directorate General for Development (DGD) for supporting the project on 'Enhancing the resilience of agro-ecosystems in Central Africa: a strategy to revitalise agriculture through the integration of natural resource management coupled with resilient germplasm and marketing approaches,' within the context of the Consortium for Improving Agriculture-based Livelihoods in Central Africa (CIALCA).

References

1. World Bank, *Agriculture for Development – World Development Report 2008*, World Bank, Washington DC, 2007.
2. P. A. Sanchez, *Properties and Management of Soils in the Tropics*, Wiley, New York, 1976.
3. FAO and World Bank, *Farming Systems and Poverty. Improving Farmers' Livelihoods in a Changing World*, FAO and World Bank, Rome and Washington DC, Italy and USA, 2001.
4. M. Kherallah, C. Delgado, E. Gabre-Madhin, N. Minot and M. Johnson, *Reforming Agricultural Markets in Africa*, International Food Policy Research Institute, Washington DC, USA, 2002.
5. P. Tittonell, B. Vanlauwe, P. Leffelaar, E. Rowe and K. E. Giller, *Agric. Ecosyst. Environ.*, 2005, **110**, 149.
6. K. Shepherd and M. J. Soule, *Agric. Ecosyst. Environ.*, 1998, **71**, 133.
7. P. Tittonell, B. Vanlauwe, P. Leffelaar, E. Rowe and K. E. Giller, *Agric. Ecosyst. Environ.*, 2005, **110**, 166.
8. B. Vanlauwe, A. Bationo, J. Chianu, K. E. Giller, R. Merckx, U. Mokwunye, O. Ohiokpehai, P. Pypers, R. Tabo, K. Shepherd, E. Smaling, P. L. Woomer and N. Sanginga, *Outlook Agric.*, 2011, **39**, 17.
9. P. A. Sanchez, in *Inaugural and State of the Art Conferences,* Transactions 15th World Congress of Soil Science, Acapulco, Mexico, 1994, pp. 65.
10. Abuja Fertilizer Summit, *Abuja Declaration on Fertilizer for African Green Revolution*, African Union Special Summit of the Heads of State and Government, Abuja, Nigeria, 2006.
11. K. A. Annan, *Forging a Uniquely African Green Revolution*, Address by Mr. Kofi A. Annan, Salzburg Global Seminars, Austria, 2008.

12. M. J. Swift, O. W. Heal and J. M. Anderson, *Decomposition in Terrestrial Ecosystems*, Studies in Ecology Volume 5, Blackwell Scientific Publications, Oxford, UK, 1979.
13. G. Cadisch and K. E. Giller, *Driven by Nature: Plant Litter Quality and Decomposition*, CAB International, Wallingford, UK, 1997.
14. C. A. Palm, C. N. Gachengo, R. J. Delve, G. Cadisch and K. E. Giller, *Agric. Ecosyst. Environ.*, 2001, **83**, 27.
15. M. Rufino, E. C. Row, R. J. Delve and K. E. Giller, *Agric. Ecosyst. Environ.*, 2006, **112**, 261.
16. A. Bationo, F. Lompo and S. Koala, *Agric. Ecosyst. Environ.*, 1998, **71**, 19.
17. B. Vanlauwe, J. Wendt and J. Diels, in *Sustaining Soil Fertility in West-Africa*, ed. G. Tian, F. Ishida and J. D. H. Keatinge, SSSA Special Publication Number 58, Madison, USA, 2001, pp. 247.
18. B. Vanlauwe, A. Bationo, R. J. Carsky, J. Diels, N. Sanginga and S. Schulz, in *Proceedings of a SoilFertNet Meeting*, 8–11 October 2002, Vumba, Zimbabwe, 2003, pp. 3.
19. E. Kiptot, Seeing beyond fertiliser trees: A case study of a community based participatory approach to agroforestry research & development in Western Kenya, PhD thesis, WUR, The Netherlands, 2007.
20. B. Jama, C. A. Palm, R. J. Buresh, A. Niang, C. Gachengo, G. Nziguheba and B. Amadala, *Agroforestry Syst.*, 2000, **49**, 201.
21. N. Sanginga, K. Dashiell, J. Diels, B. Vanlauwe, O. Lyasse, R. J. Carsky, S. Tarawali, B. Asafo-Adjei, A. Menkir, S. Schulz, B. B. Singh, D. Chikoye, D. Keatinge, and O. Rodomiro, *Agric. Ecosyst. Environ.*, 2003, **100**, 305.
22. B. Vanlauwe, P. Pypers, E. Birachi, M. Nyagaya, B. van Schagen, J. Huising, E. Ouma, G. Blomme and P. van Asten, in *An Eco-Efficiency Revolution in Tropical Agriculture: Climate-Smart Production Systems for Smallholders*, CIAT, Cali, Colombia, in press.
23. B. Vanlauwe, K. Aihou, S. Aman, E. N. O. Iwuafor, B. K. Tossah, J. Diels, N. Sanginga, R. Merckx and S. Deckers, *Agronom. J.*, 2001, **93**, 1191.
24. R. Gentile, B. Vanlauwe, P. Chivenge and J. Six, *Soil Biol. Biochem.*, 2008, **40**, 2375.
25. B. Vanlauwe, J. Kihara, P. Chivenge, P. Pypers, R. Coe and J. Six, *Plant Soil*, 2011, **339**, 35.
26. G. Nziguheba, R. Merckx, C. A. Palm and R. M. Rao, *Biol. Fertility Soils*, 2000, 32, 328.
27. K. E. Giller, E. Witter, M. Corbeels and P. Tittonell, *Field Crops Res.*, 2009, **114**, 23.
28. H. G. Buffett, *Nature*, 2012, **484**, 455.
29. B. Vanlauwe, J. Diels, N. Sanginga and R. Merckx, *Plant Soil*, 2005, **273**, 337.
30. B. Vanlauwe, J. Diels, K. Aihou, E. N. O Iwuafor, O. Lyasse, N. Sanginga and R. Merckx, in *Integrated Plant Nutrient Management in sub-Saharan Africa: From Concept to Practice*, ed. B. Vanlauwe, J. Diels, N. Sanginga and R. Merckx, CABI, Wallingford, UK, 2002, pp. 173.

31. B. Vanlauwe, K. Aihou, B. K. Tossah, J. Diels, N. Sanginga and R. Merckx, *Plant Soil*, 2005, **269**, 285.
32. K. Aihou, N. Sanginga, B. Vanlauwe, O. Lyasse, J. Diels and R. Merckx, *Agroforestry Syst.*, 1999, **42**, 213.
33. P. Tittonell, K. Shepherd, B. Vanlauwe and K. E. Giller, *Agric. Ecosyst. Environ.*, 2007, **123**, 137.
34. S. N. Guto, P. Pypers, B. Vanlauwe, N. de Ridder and K. E. Giller, *Field Crops Res.*, 2011, **122**, 85.
35. S. Zingore, H. K. Murwira, R. J. Delve and K. E. Giller, *Field Crops Res.*, 2007, **101**, 296.
36. R. Gentile, B. Vanlauwe and J. Six, *Ecol. Applicat.*, 2012, **21**, 695.
37. B. Vanlauwe, N. Sanginga, K. Giller and R. Merckx, in *Agriculture and the Nitrogen Cycle*, ed. A. R. Mosier, J. K. Syers and J. R. Freney, SCOPE Nr 65, Island Press, Washington, USA, 2004, pp. 115.

Climate Change and Forest Dynamics: A Soils Perspective

WENDY PETERMAN* AND DOMINIQUE BACHELET

Conservation Biology Institute, 136 SW Washington Avenue, Suite 202, Corvallis, OR 97330, USA
*E-mail: wendy@consbio.org

ABSTRACT

Increasing temperatures have been recorded around the world, leading to changes in precipitation, sea-level rise and extreme events. Climate models are currently in use to simulate the effects of these changes on vegetation cover, which is a strong indicator of ecosystem changes in response to various drivers. Climate change, as well as anthropogenic stressors, is affecting forest dieback and tree-species migration. This chapter addresses the connections between changes in various forest types and the global soil carbon, nitrogen and hydrologic cycles, and related feedbacks between these factors and both natural and anthropogenic environmental changes. We discuss the ways these feedbacks between land use, vegetation changes and global nutrient and water cycles can lead to further climate change and soil degradation, which have profound effects on food security, and we conclude by proposing the use of soil characteristics as tools to inform land managers of challenges they may face in preserving valuable services from forested lands and cropping systems.

Issues in Environmental Science and Technology No. 35
Soils and Food Security
Edited by R.E. Hester and R.M. Harrison

Published by the Royal Society of Chemistry, www.rsc.org

1 Introduction

Widespread forest mortality is a worldwide phenomenon, and scientists researching possible causes often arrive at the conclusion that effects of climate change are behind much of the forest dieback.[1,2] In the past two decades, warmer temperatures and decreased precipitation have been identified as the main causes for pest outbreaks, increased forest fire frequency and extended drought-related stress.[3,4]

Climate change is an important factor leading to forest dieback and tree species migration as they relate to drought, water stress, early snow melt, reduced snow cover, pest outbreaks and fire risk.[1,2,5,6] Forest ecosystems are facing many stresses, both natural and human-caused that can contribute to changes in forest dynamics. Anthropogenic stressors are often related to land conversions for agriculture or urbanisation, fire suppression or initiation, and pollution. Natural stressors may include severe drought, waterlogging and cold, and secondary insect attacks and diseases of stem and root fungi.[7] Although some forests might have a positive response to increased carbon dioxide emissions and longer growing seasons,[8,9] this response appears to be regional or temporary as current global forest loss is exceeding forest gain[10] (Figure 1).

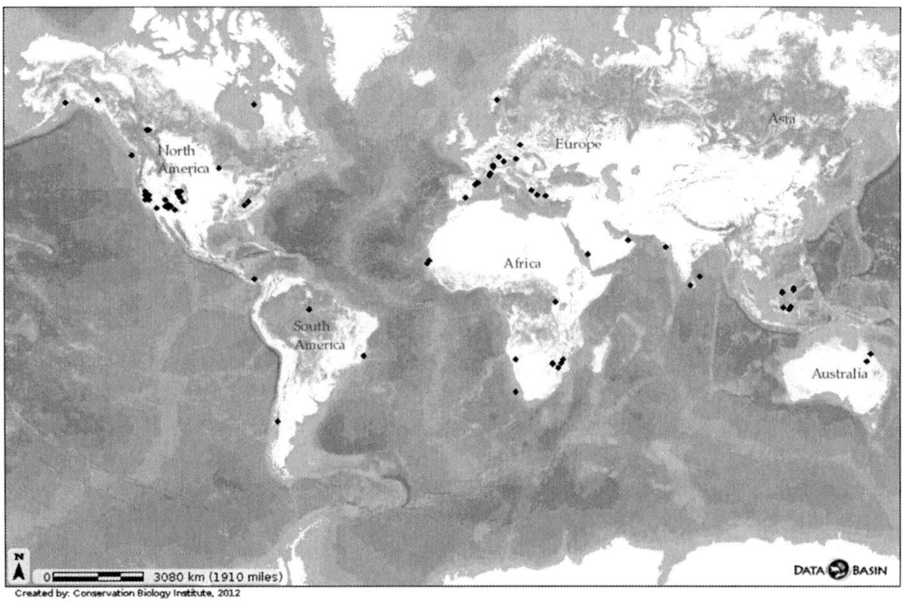

Figure 1 This map shows the locations of forest dieback documented in a 2010 publication.[2] An interactive version of this map, including details of forest type, dieback causes and extents and original data sources can be found at: http://app.databasin.org/app/pages/datasetPage.jsp?id=b2947eeae2e5488a86eacf0fcd4df7a4 (Source: Dr. Joerg Steinkamp, Biodiversity and Climate Research Centre, Wendy Peterman, Conservation Biology Institute).[11]

Due to limitations in the understanding of forest physiology, climate and mortality, forest die-offs are a big uncertainty in climate projections of terrestrial ecosystem impacts, climate/ecosystem interactions and carbon-cycle feedbacks.[12] Scientists are always seeking greater understanding of the complex mechanisms leading to forest dieback, migration and shifts in species dominance to help predict where and when these changes may occur.

Soils hold important clues about shifts in hydrology and vegetation across the landscape because, in terrestrial systems, soil characteristics govern the reception, storage and redistribution of precipitation. This, in turn, determines the supply of plant-available water and, indirectly, the nutrients necessary for plant establishment and growth. Because soils with more water are less sensitive to warming, changes in soil moisture result in changes in soil heat capacity and conductivity, which, in turn, affect infiltration and water transport in the soil profile.[13,14] Soil response to changes in precipitation has implications for vegetation water needs, fire risk, pest outbreaks, infiltration rates and groundwater recharge;[15] therefore, in-depth analyses of these soil characteristics can give scientists and managers the tools they need to predict where trees will be most vulnerable to future water stress and where they will be most likely to establish and thrive under future conditions.

In this chapter, we review existing literature for examples of on-going forest responses to climate change, many of which are also exacerbated by anthropogenic stressors. We discuss the implications for the global carbon, nitrogen and hydrological cycles and how resulting changes in forest lands and associated ecosystem services may affect food security in the future. We conclude by providing a method to use soil characteristics to inform land managers of challenges they may face in preserving valuable services from forested lands and cropping systems.

2 Projected Trends in Climate Change

2.1 General Climate Trends

Rising temperatures have been recorded around the world and are projected to continue to rise, with regional and local patchiness, causing an overall decrease in the longevity, extent, and thickness of glaciers, ice sheets, and snowpacks.[16] Observations have shown that land has been warming up at a faster rate than oceans, due to the greater inertia of deep oceans. Sea-level rise projections presented in the last IPCC report[2] were extremely conservative, and new publications suggest that higher levels are likely to be reached by the end of the century, if current trends of ice-melting and ocean-warming continue.[17]

Global warming is likely to drive an increase in global mean precipitation (rain and snowfall). However, the degree of spatial and seasonal variation remains large, even when considering multi-model means. All simulations point to increases in precipitation at high latitudes where more rain than snow has recently been observed, a trend that probably will continue as winter temperatures increase. There is also a general agreement over precipitation decreases in the sub-tropics. Models agree to a lesser extent over an increase of precipitation in the tropics, and cloud formation and wind patterns are areas of uncertainty in model structure, as current understanding remains limited.

Natural climate variability (*e.g.* El Niño Southern Oscillation (ENSO), Pacific Decadal Oscillation (PDO), Atlantic Multi-decadal Oscillation (AMO)) and its impacts have been well documented for many regions of the world,[18] but the understanding of the causes of shifts in teleconnections (related climate anomalies) remains limited and thus difficult to include in climate models.

Extreme events (long, intense droughts, flood, hurricanes and typhoons) are also difficult to predict from general circulation models. The latest report from the Intergovernmental Panel on Climate Change (IPCC)[19] warns about the increased risk of more intense, more frequent and longer-lasting heatwaves, as exemplified by the European heatwave of 2003 that killed several thousands of people and caused widespread forest mortality.[20] Along with a greater risk of drought, there is an increased chance of intense precipitation and flooding due to the greater water-holding capacity of a warmer atmosphere, such that both wet and dry extremes should become more severe. Several modelling studies are projecting that future cyclones could become less numerous but more severe, with greater wind speeds, more intense precipitation, and higher ocean waves.

These extreme events, while unpredictable, often shape our landscapes. Past extreme events such as the drought of the 1930s that caused the Dust Bowl in the USA, or the 1998 floods in China caused by heavy rainfall that affected 240 million people, certainly affected natural ecosystems and human land use. Recently, reports of extreme events have been increasing. For example, a drought in the of summer 2010 caused crop failure and huge fires in Russia, while record rainfall caused extensive flooding and loss of life in both China and Pakistan. These extremes might be evidence of climate destabilisation, but they are at the very least consistent with what climate scientists have been expecting. They certainly pose a challenge to the more comfortable prospect of chronic linear change rather than abrupt and unpredictable change, yet these events might be what people most need to take into account when they consider preparing for change. In the past, the reliability of models was tested in part by simulating large disturbances and observing the simulated system's response. It may be necessary for practitioners to focus on disturbance simulation to fully explore the resilience of their systems.

3 Changes in Forest Dynamics

3.1 Introduction

Forests and woodlands account for approximately 30% of terrestrial land cover[21] and store about 45% (more than 1 trillion tonnes) of the carbon in terrestrial ecosystems.[12,22] Changes in vegetation cover are strong indicators of ecosystem changes in response to many change agents, including land use and climate change. Forests can be either contributors or inhibitors of climate change on regional scales, but they have the potential to play a significant role in mitigating the pressures of global environmental change.[22]

Land-use practices and climate change may work in concert to weaken the ability of trees to defend themselves against pests, improve conditions for native pests to flourish, and introduce unfamiliar pests to new locations. Non-native forest pests have increased globally since the 1990s due to increased international trade and other human activities[23] and now threaten forest productivity and diversity.[24–27] For example, the root fungus *Phytophthora cinnamomi*, which originated in Papua New Guinea and is now contributing to forest mortality in the USA and Australia, and to Iberian oak decline in the Mediterranean and coastal northwestern regions of Europe,[7] is a pathogen requiring warm, wet soils to infect roots, and extreme weather conditions, such as drought or waterlogging, can increase the susceptibility of trees to infection.[7]

In addition to non-native pest introduction, pollution is another human cause of altered forest dynamics around the world. Forests adjacent to urban and agricultural areas are responding to increased nitrogen deposition and other airborne pollutants.[28,29] Human activities such as road-building and land conversion lead to landscape fragmentation, which increases forest edges.[29] These edges can be "hotspots" of dry deposition, with as much as four times the rate of atmospheric nutrient delivery as areas without edges.[30,31] The difference between pollutant concentrations from the forest edge to the interior can be very large, possibly even exponential, especially when particles are transported horizontally by wind.[29] Several studies of excessive nitrogen deposition have shown that the cumulative effects of nitrogen additions over many years can be negative due to a phenomenon called "nitrogen saturation",[32] which can ultimately lead to nitrogen leaching into surface waters.[33]

3.1.1 Tropical Peat Swamp Forests

Montane peat swamps in cloud and other tropical forests play a significant role in the global carbon cycle as they store a considerable amount of carbon in their soils.[34,35] Approximately 60% of the known peatland forests are in south-east Asia. Peat soils form from decayed woody plant debris decomposing in high precipitation and temperature conditions in swamp forests at low elevations in river valleys.[36,37]

A high estimate of the remaining historical peat swamp forests is 36%.[38] Drying of peat swamps through logging or for agricultural use is increasingly common, but when these soils dry, they are extremely flammable.[39–41] Peat soils are unique in their ability to burn above and below ground.[42] Clearance and burning of peat swamp forest in south-east Asia could contribute to 3% of total global human emissions.[43,44] The 1998 Indonesia fires burnt some 8 million hectares of land and, according to scientific estimates, released between 0.48 and 2.57 Gt of CO_2 into the atmosphere, which is between 13 and 40% of the mean annual global carbon emissions from fossil fuels. The exact amount remains uncertain.[45]

3.1.2 Tropical Rainforests

Changes in Amazon rainforest ecosystems have the potential to affect not only the global carbon budget, but the hydrological cycle and feedback to global climate as well. Climate and air quality in the Amazon region are highly dependent on feedbacks between vegetation cover, land surface and biogeochemical fluxes.[46] Approximately eight tonnes per year of water evaporates from Amazon forests.[19] Run-off from the Amazon basin to the Atlantic Ocean accounts for 15–20% of the global freshwater flow to oceans.[47] Amazon forests also contain currently between 90 and 140 billion tonnes of carbon,[48] which is about nine-to-fourteen decades of the current anthropogenic carbon emissions.[49]

Some simulations indicate Amazon forests will convert to grasslands by the end of the 21st century,[50–54] causing dramatic changes in soil and hydrologic conditions.[55] The HadCM3 general circulation model projects that regional warming and drying will also lead to large-scale forest dieback.[56] Grasses can expand into disturbed forest patches *via* animal or wind seed dispersal, but more often there is deliberate grass seeding by ranchers after logging.[57] As of 2001, deforestation in the Amazon had reduced the original forest area from 6.2 million km^2 to 5.4 million km^2 (*i.e.* 87% of original area),[56] and existing plans to build new infrastructure could further reduce Amazon forests to 3.2 million km^2 (53% of original) by 2050.[48]

The overall indications of research on Amazon forest decline is that forest degradation by humans or climate could lead to an even hotter, drier climate in Amazonia. Deforestation reduces the recycling of water,[44,55] while decreased forest evapo-transpiration leads to decreased surface cooling, which leads to warmer air temperatures, higher evaporative demand and increased water stress.[55,56] Fire hazard also increases in a drier climate, potentially causing an increase in smoke and dust aerosols that could alter the frequency and amount of precipitation received.[55,58] A coupled climate-carbon model from the UK Meteorological Office Hadley Centre showed that severe drying of Amazon forests would lead to forest losses, resulting in feedbacks at both regional and global scales, further magnifying drought conditions and forest degradation.[46,56,59]

Amazon forests on dry margins or on shallow, infertile soils are most vulnerable to drying.[55] Amazon trees avoid drought stress by penetrating deeply into the soil to access deep soil water, and they utilise hydraulic redistribution of water to more shallow soil horizons.[60–64] The threshold of drought tolerance in Amazon forests has been shown experimentally as an available soil water capacity less than 30% of its maximum value.[62,65] During the El Niño 1997–1998, forest mortality increased 50% post-drought[66] as canopy dieback increased radiant energy into the forest and increased the temperature in the forest interior, further drying the soils and increasing fire risk.[57,67]

In addition to climate change, human pressure on Amazonia is important. Humans look to this region to exploit biofuels to substitute for oil, cattle and swine industries, agro-industry expansion, sugar cane for ethanol, palm oil for biodiesel and soy crops.[55–57] Land-use changes increase habitat fragmentation, edge effects of pollution and dry air circulation under forest canopy, and fire ignition sources.[57]

3.1.3 Temperate Forests

In the USA, the regional importance of many tree species is changing rapidly. Some tree species are experiencing dieback in response to precipitation and temperature changes, while others are seeing shifts in species dominance. In a regression tree analysis of eighty common trees species under five future climate scenarios for 2100 counties in the eastern USA, Iverson and Prasad[68] project that average species richness may remain the same or even increase with climate change, but there are likely to be dramatic changes in forest type in this region.

All five models[68] predict an extirpation of spruce-fir forests in New England, USA, and all but the two least-severe models show an extirpation of aspen and birch species (both still largely reduced). Maple, beech and birch species are largely reduced under all scenarios. The main increases are seen in oak-hickory and oak-pine woodlands, which are projected to increase 34% and 290%, respectively. The loblolly shortleaf pine is projected to decrease by 32% and shift its range to the north and west. Longleaf slash pine is projected to decrease by 31%, but elm-ash-cottonwood woodlands are projected to remain in the upper Great Plains region of the USA.

According to their consensus models, 24 species in the eastern USA will see a decline of at least 10%, while 35 species will see an increase in regional importance, with 12 of these increasing by 100% or more. Recent national forest assessments[21] indicated that the total area of USA national forests has been increasing annually, but that the amount of increase is slowing dramatically. More recent US Forest Inventory Analysis (FIA) studies[69] show that some eastern states are levelling off in forest increase and others are beginning to decline.

Coops and Waring[70,71] used the 3-PG process-based model to predict forest responses to climate change. Predictions of future tree distributions in the Pacific Northwest of the USA show large changes in lodgepole pine and ponderosa pine distributions. They show that lodgepole pine is most likely to persist at sites with significant spring frost, summer temperatures below 15 °C, and soils that are fully recharged from snow melt. Using future climate projections, they predict a decrease of 8% in suitable lodgepole pine habitat and, by the last 30 years of the 21st century, they predict that the species will be absent from most of its current range. By 2050, there is likely to be a significant decrease in its distribution, especially in central Oregon and Washington, British Columbia and the western side of the Rocky Mountains, so that by 2080 lodgepole pine is projected to be gone from Oregon, Washington and Idaho (Figure 2). The most restricting factor in the pine persistence is soil water.

As the climate warms, tree species are expected to migrate to higher altitudes and higher latitudes into areas previously characterised by low temperatures. The compositions of high-elevation forests are changing rapidly.[15] Altitudinal tree lines are seen as the most sensitive to global warming, because historic temperature decreases at higher altitudes have been the main limitation on tree lines globally.[73] As temperatures increase, altitude becomes a less reliable predictor of tree-line limitation. In the Andes, Chile, Patagonia and the Rocky Mountains of Montana, tree growth at tree lines did increase in brief pulses with subsequent infilling over several centuries, but, in the past fifty years,

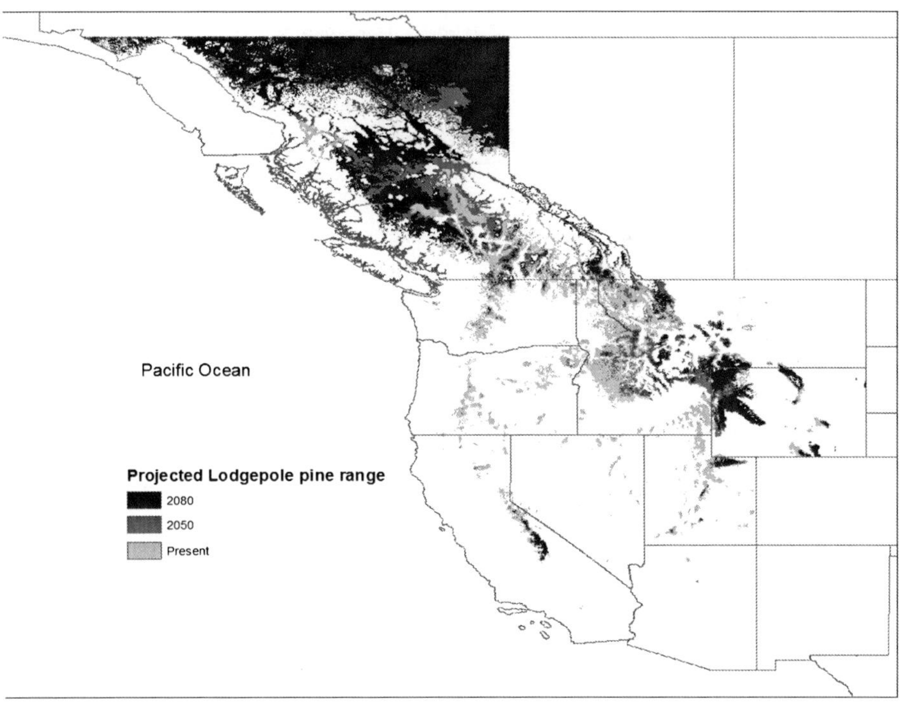

Figure 2 This map shows the lodgepole pine (*Pinus contorta*) projected range contraction in North America between the present and 2080.[72] (An interactive version of this map and other maps of future species distribution simulations can be found at: http://app.databasin.org/app/pages/galleryPage.jsp?id=896ee1c381fd4a50b5f811b4b11c0898).

these regions have shown actual declines at the tree line, with the biggest decline 200 m below the tree line.[73]

Although many species have been observed to move uphill in response to temperature, this change alone is not enough to understand future plant distributions. There is evidence of some downhill shifts despite climate warming.[74–76] Drivers of species distribution shifts and mechanisms are not well understood at present, but it's possible that some species may track decreases in water deficit to lower elevations rather than temperature increases to higher elevations.[77] In the past century there has been an increased density of younger tree cohorts at lower elevations.[78,79]

3.1.4 Alpine Forests

Climate projections show the greatest and earliest warming trends at higher latitudes (45 to 65 °N), especially in continental interiors.[80,81] At the thermal ecotone, small changes in temperature can have large consequences.[82] The interior regions of Alaska are good places to investigate possible climate-change effects on vegetation composition and soil thermal dynamics, because changes in snowmelt affect surface and sub-surface soil moisture through interactions with permafrost.[83] Changes in surface hydrology and soil temperature affect forest and tundra vegetation as well as forage availability for ungulates.[84] In Alaska, the tree line is shifting northward into the tundra, important migratory bird and caribou habitat.[15]

Yellow cedar (*Callitropsis nootkatensis*), a tree found in British Columbia, Canada, and the Pacific Northwest of the USA, is experiencing varied responses to changes in soil temperature and hydrological conditions.[85] Between the 19th century and the present, yellow cedar has seen a high rate of mortality only in SE Alaska and bordering regions of Canada, with dieback symptoms originating in the roots, spreading to the crown and finally manifesting in the bole of the trees.[86] The emergence of the initial symptoms in the roots indicates soil factors in the causes of stress,[85] which generally makes trees more susceptible to predators; however, biotic agents are not the primary cause of death.[87–93]

Yellow cedar is usually drought tolerant and successful in poor soils,[94] but it also tends to decline on poorly-drained soils,[87] and in the late 19th century a warming period at high latitudes[95] was associated with yellow cedar decline.[96] D'Amore and Hennon[87] explored the soil conditions connected to cedar decline and concluded that warmer air and soil conditions, reduced snow packs, and early spring warming are causing trees to de-harden too early and to become susceptible to late-frost injury.

3.2 Hydrologic Responses

Changes in temperature, precipitation and vegetation cover have major implications for the global water balance in which soils play an integral role.

One of the most valuable ecosystem services provided by forests at the watershed scale is to provide clean drinking water. With warming, there will be an increased need for forest cover to provide shade and reduce evaporation, protecting soils and streams from water losses, while potential declines in their extent due to human land-use or increased climate-driven mortality may reduce their ability to do so.[97] Although some studies have shown that forest harvest on certain soils with certain topography can increase water yields, particularly when precipitation exceeds evapotranspiration, this increase in water from vegetation removal is small and short-lived,[98] and it has the potential to increase the frequency of landslides, because root removal destabilises soils.

Furthermore, small changes in the proportions of winter precipitation *versus* snow can greatly alter seasonal stream flow throughout the year.[99] Simulations run by Tang and Zhuang[100] show that increased precipitation combined with earlier snowmelt and delayed snow onset, lead to longer snow-free periods in tundra and boreal forests, which could increase the growing season by up to three weeks during the 21st century. In Alaska, USA, higher temperatures over the last century have led to changes in the length of the growing season for terrestrial ecosystems,[101] and increased evapotranspiration in response to warming is also leading to an overall decline in spring soil moisture.[102]

In southeast Asia, "cloud" forests that take water from clouds to augment ground water and mountain streams, sometimes doubling the effective rainfall in the dry season and increasing total forest moisture inputs by 10%,[103,104] are threatened by climate change as well as expanding infrastructure, forestry and agriculture. Currently, tropical montane forests cover about 92 million hectares (15% of tropical rain forests).[105] One half of the tropical montane forests can be found in south-east Asia (approximately 32 million hectares).[106] The biodiversity of native frogs, birds and mammals can be higher than in lowland tropical rain forests, making these habitats critical for conservation.[107,108]

The cloud forests of Malaysia are declining 23% faster than lowland forests in the region.[109] Clearing for agriculture and cattle leads to pesticide and fertiliser contamination in surrounding watersheds, decreased water yields in highland streams,[110] and severe soil erosion and stream sedimentation.[111] In addition to land conversion, commercial selective logging currently affects 1.1% of cloud forests globally (higher than other tropical forests).[112] Possible solutions to cloud-forest decline may be increasing protected areas to preserve intact ecosystems and integrating the forests into desired human uses through agroforestry and increased animal husbandry, and thus increasing land-use efficiency.[104]

3.3 Carbon Responses

One of the most important ecosystem services that forests deliver is in their role as a large carbon sink. According to the US National Climate Assessment,[113] if one-third of the current croplands were converted to forests, the US carbon emissions could be reduced by as much as 10%. Carbon uptake depends on climate, disturbance and management legacy, age and type of the forests.[114] While water availability limits productivity in semi-arid grass, shrub, woodland and dry forest ecosystems,[115] at high latitudes where low temperatures limit water availability, early snowmelt and soil thaw initiate photosynthetic carbon uptake, and warming induces a longer growing season where radiation is the limiting factor.[116] Soil properties limit the availability of both water and nutrients, potentially limiting plant growth. Soil texture, depth, salinity and topography have strong local influences on forest growth.[117–121]

Soil organic carbon (SOC) pools may store as much as 90% of the carbon in terrestrial ecosystems[122] but can also vary widely in response to woody plant encroachment (from -6200 g C m^{-2} to +2700 g C m^{-2}).[121,123–131] The increase in woody encroachment of grasslands and deserts during the 20th century has been attributed to a variety of factors, such as increased atmospheric CO_2 concentration, land-use change (grazing), climate patterns and fire suppression.[132–134]

Woody plants influence SOC pool sizes, particularly beneath their canopies,[135] through litter accumulation; they can affect soil respiration and leaching (roots and microbes) and reduce erosion.[121] Because of the greater rooting depths and higher root lignin content of woody plant species, soil carbon is generally higher in shrublands than grasslands[122,136,137] and is accompanied by more resistant organic matter in deeper soil layers.[138–141] Barger *et al.*[121] showed that bulk density and clay content mediate the magnitude and direction of SOC changes with woody encroachment. Increases in bulk density are linked to low SOC, and carbon losses are associated with soils of bulk densities greater than 1.6 g m^{-3}. In the southern Great Plains, USA, SOC accumulation rates are three times greater in fine soils than in adjacent coarse soils,[136] and woody encroachment with higher SOC contents has been associated with a clay gradient.[142–145]

3.4 Nitrogen Responses

Nitrogen availability is closely tied to the water cycle[146,147] and it controls photosynthetic rates and thus forest productivity, as well as carbon allocation and resulting canopy development.[148–156] Nitrogen can limit carbon uptake even when water is readily available, but when water is limiting, plants cannot take up available nitrogen unless they develop a symbiotic relationships with a nitrogen fixer.[157–159] For this reason, the sizes of the soil carbon and nitrogen pools are good indicators of any change in the local soil nitrogen-supplying capacity.[160–162]

Nitrogen dynamics are very much driven by the constant feedbacks from plant, soil and microbial interactions.[163] Tree species influence nitrogen cycles in different ways through root uptake, mycorrhizal associations, exudation and the chemical quality of plant litter,[164] and trees in the same climate with different soil fertility can exhibit different rates of growth and above- and below-ground nitrogen accumulation patterns.[165] For example, root turnover and exudation provide a large carbon and nutrient source for soil microbial communities[166] which can generate rhizosheaths to enhance plant nutrient and water uptake by creating a beneficial microenvironment around the roots. Mycorrhizal hyphae can also allow plants better access to resources in the various soil layers beyond the tree canopy.

The age and land-use history of a forest determines overall nitrogen availability. Older, aggrading systems retain their nitrogen biomass in the soil, but disturbed systems lose nitrogen in large pulses, decreasing nitrogen mineralisation for plant uptake.[167–169] When fires burn forest litter and understory, immobilised nitrogen in the biomass gets released to the atmosphere.[170,171] Other forest nitrogen outputs include biomass loss from harvest, erosion, leaching and gaseous transfers.[172]

Forests in humid temperate ecosystems are historically nitrogen limited, but Skeffington and Wilson[32] published a new theory of forest "nitrogen saturation" in response to increased levels of atmospheric nitrogen deposition. When background nitrogen levels are low, temperate forests usually experience sub-optimal nitrogen availability, and nitrogen additions can enhance tree growth on very short-term intervals.[173,174] Under conditions of elevated nitrogen deposition, there is the potential for forest nitrogen concentrations to exceed plant and soil uptake, leading to nitrogen losses from the system.[33,163] Nitrogen saturation is characterised by increased nitrate losses from forest soils in spring snowmelt and soil water percolating below the rooting zone during the growing season.[33,175] As negatively charged nitrate ions leave the soil, they combine with positively charged ions such as calcium and aluminum that leach as well, causing decreased soil fertility and increased acidity.[33] Consequently, excessive nitrogen concentrations add stress to forest ecosystems in temperate regions and may lead to decreased forest production and eventually decline, as nitrogen-saturated forests become net nitrogen sources rather than sinks.[33,163] Furthermore, leached nitrate reaching streams affects water quality and has implications for nitrous oxide emissions to the atmosphere.

Nitrogen leaching is highly dependent on precipitation and snowmelt as well as the amount of water infiltrating below the rooting zone in the soil.[176,177]

At the Hubbard Brook Experimental Forest in New Hampshire, USA, Bernal *et al.*[178] observed that a decline in snowpack is turning the soil into a massive nitrate sink. Repeated model simulations of the system suggest that two mechanisms of soil organic nitrogen storage are responsible for the observed decrease in nitrate export: (1) more nitrogen is being held in the soil as decreased snowpack reduces the water flow paths, allowing more opportunities for microbes and plant roots to immobilise nitrates, and (2) the gradual accumulation of nitrogen as the forest recovers from abrupt nitrogen losses due to past timber harvests, hurricanes or ice storms. There also appears to be a small effect of the recent sugar maple decline on the change in nitrate export.

4 Food Security Implications of Forest and Soil Responses to Global Change

4.1 Anthropogenic Soil Degradation

Forest degradation and the resulting soil degradation are closely tied to the major issues facing world food production in the face of global environmental change. Because soil degradation is affecting crop productivity and contributing to malnourishment around the world,[179] improving soil quality is essential to maintaining life on earth.[180] Soil health is a high priority listed by the United Nations' Millennium Project hunger task force and the United States Department of Agriculture reports that decreasing degraded soils and increasing crop yields by 0.1% could reduce the number of starving people by 5% in a decade. Global hotspots of degradation include central and southern Asia, China, the Andes, the Caribbean, and the savannas of South America.[181]

In 1991–1992, the International Soil Reference and Information Center (ISRIC) developed a global database of human-induced soil degradation. Soil degradation derives from increasing pressure on land to improve living conditions, provide higher standards of living, or simply allow human survival. Five human causes of soil degradation are: (1) deforestation or removal of natural vegetation for agricultural use, roads, timber harvest or urbanisation, (2) overgrazing, (3) inefficient agricultural practices, (4) overexploitation of vegetation for domestic use, such as fuels and fencing (incomplete vegetation removal is insufficient to prevent topsoil removal), and (5) bio-industrial activities that lead to soil pollution.[181]

In a 2004 *Science* article, J. Kaiser[179] gave an overview of global soil degradation and its impacts on regional food scarcity. Soil degradation is given as the main obstacle to reducing hunger in Africa and the cause of the current devastation in Haiti. The previously forested landscape of Haiti has been severely denuded until only 3% of the original forest cover remains. At least one-third of the landscape has lost too much topsoil to be able to support crops. In the 1930s, the USA temporarily experienced a similar plight when a combination of poor land-management and drought contributed to a massive loss of topsoil in the midwestern states. In China, the Loess Plateau, the site of the fastest topsoil loss in the world, loses approximately 1.6 tons of loess each year, and some lands in the lower Himalayas have totally lost the capacity for food production. In sub-Saharan Africa, where farmers cannot afford fertilisers, and crop residues and animal excrement are used for fuel, soil fertility is quickly declining. In some parts of Africa, farmers traditionally rested fields, but now land constraints are too tight. In the Middle East and India, poor irrigation is leading to salinisation of soil, and in Australia the eradication of some native plants is causing dramatic changes in local water tables, leading to salinisation of the topsoil. Desert expansion, driven by conversion of grasslands in the Sahel of Africa, Kazakhstan, Uzbekistan and northern China, has led to wind erosion and dust storms.

4.1.1 Mechanisms of Soil Degradation

Oldeman[181] defined two categories of soil degradation: (1) the displacement of soil material by water or wind erosion, and (2) *in situ* soil deterioration through chemical or physical processes. Erosion or topsoil removal reduces soil fertility and may reduce crop rooting depths. Deforestation, overgrazing, and agriculture are the main causes of water erosion, because they expose soil to the direct impacts of rainfall, and wind erosion is almost always caused by a decrease in vegetation cover from overgrazing or the removal of vegetation for another use. Chemical degradation can be from loss of nutrients and organic matter (insufficient fertilisers, using poor soils, removal of natural vegetation), salinisation (from poor irrigation practices), acidification (from over application of fertilisers), or pollution (in industrialised nations with high population densities). Physical degradation includes compaction or sealing (from heavy machinery, low organic matter or high silt), waterlogging (from human intervention in natural drainage systems), or the subsidence of organic soils (drainage or oxidation of peat soils).

Soil chemical and physical weathering rates are driven by vegetation, temperature, and precipitation.[181,183] This can be greatly intensified in areas where land use accelerates soil denudation, exposing more mineral surface area and rocks.[184] Bayon *et al.*[185] showed that chemical weathering of surface minerals due to intensified human land-use and forest clearing, rather than regional climate change, may have led to an abrupt vegetation shift from rainforest trees to savannas in Central Africa 3000 years ago. Records of past vegetation patterns show a great loss of primary forests as they were replaced by savannas and other pioneer species between 3000 and 2200 ago. At the same time, archaeological research shows that Bantu-speaking people migrated into the region and cleared forest for agriculture and iron smelting.[186–191] This large-scale deforestation event may still influence current vegetation patterns in African rainforests.[192,193]

4.1.2 Soil Degradation Implications for Soil Carbon and Nitrogen

Land erosion plays a significant role in global nutrient cycles. Soil organic carbon and soil nitrogen are both easily removed by wind and water erosion, which can lead to feedbacks to the atmosphere. Land cultivation leads to organic matter losses, directly affecting the soil chemical, physical and biological properties that affect crop production.[194] Khormali[195] showed that soil organic carbon and nitrogen in Iran are significantly depleted by increased water erosion from past deforestation. Vagen *et al.*[196] showed that some landscapes under cultivation more than fifty years without organic matter enrichment have extremely low organic carbon and total nitrogen content. Soil carbon and nitrogen content of the 0–10 cm layer is lower after 53 years of cultivation than it is in nearby natural forests.[197] Cultivation also reduces soil aggregate stability,[198] exemplified by the fact that average bulk density of a rainforest soil is lower than in deforested areas.[199] These reductions in soil porosity lead to soil degradation due to changes in water infiltration and percolation.[200]

4.1.3 Repairing Soil Degradation

The challenge the world now faces is how to manage forests and agricultural lands proactively to deliver food and water to humans, while also preserving biodiversity and other ecosystem services.[201] Strategies to reduce or even reverse soil degradation include no-till farming, water conservation and harvest, cover cropping, woodland regeneration, agroforestry, improved grazing practices, more efficient irrigation and erosion control.[181,179] Soil organic-matter content is of the utmost importance and can be improved by management practices that add biomass to the soil, reduce disturbance and improve soil structure.

4.1.4 The Agroforestry Alternative

From Conte,[202] "A forest understood as an agrarian landscape can include many centuries of forest-based husbandry."[203,204] For centuries, farmers in the eastern Arc Mountains of Africa used agroforestry in the mountains to cultivate native and introduced plants.[205–209] At Mt. Kisagau, traditional botanical knowledge is tightly connected to a 1000 m elevation gradient and conveyed to generations through oral history that details the agro-ecological use of the entire mountain.[202] Farming once entailed a mosaic of forest ecosystems at varying stages of exploitation and regeneration that emphasised mobility, since water rather than temperature was a key factor of tree migration across Africa.[210] Farmers combined imported grain species with African beans, sorghum and millet,[211,212] and there is evidence that Asian banana was possibly a feature in African agroforestry for more than 5000 years.[212] More recently, western-style agricultural and forestry practices ignored lessons of indigenous land-use and forest evolution, emphasising timber yields only, and the landscape was quickly degraded to a point where indigenous trees could not even be replanted.[202] Agroforestry practices have been common for thousands of years in Europe, Africa and South America, and may hold the key to understanding how to use fertile forest soils to support all life without denuding and degrading them.

5 Soil Characteristics as Tools for Adaptive Management

Scientists are currently developing new tools based on soil characteristics to help farmers and land managers evaluate the potential effects of climate change on soil water availability and develop appropriate strategies to adapt to the change or possibly mitigate negative climate impacts. Correlation models can incorporate the effects of geology, elevation and specific soil properties into a vulnerability index, estimating where crop or tree mortality will most probably occur during periods of prolonged drought or flood. With some awareness of this vulnerability, managers can implement practices to reduce soil erosion or decrease competition for scarce resources in areas at high risk of mortality, or they can focus soil quality restoration efforts in areas where forest or crop resilience is expected.

Soil physical characteristics are reliable predictors of forest health as the climate changes, because the temporal scale at which climate affects soil development is much longer (thousands to millions of years) than the scale at which climate is affecting trees (days, months or years). Soil characteristics constrain water and nutrient availability to forests and crops alike, and hold clues about how water might be moving through the soil. These factors can be indicators of whether rainfall is likely to evaporate or infiltrate, as well as the amount and duration of water storage in the rooting zone. Because they hold or release moisture based on their texture, depth and chemistry, soils can either mitigate or exacerbate climate change impacts to plants, affecting ecosystem vulnerability to heatwaves, wildfires and pest outbreaks. Therefore, soil characteristics hold the key for farmers and land managers seeking sustainable means to meet the food and energy demands of a growing population.

References

1. P. J. van Mantgem, N. L. Stephenson, J. C. Byrne, L. D. Daniels, J. F. Franklin, P. Z. Fulé, M. E. Harmon, A. J. Larson, J. M. Smith, A. H. Taylor and T. T. Veblen, *Science*, 2009, **323**, 521–524.
2. C. D. Allen, A. K. Macalady, H. Chenchouni, D. Bachelet, N. McDowell, M. Vennetier, T. Kitzberger, A. Rigling, D. D. Breshears, E. H. Hogg, P. Gonzalez, R. Fensham, Z. Zhang, J. Castro, N. Demidova, J. -H. Lim, G. Allard, S. W. Running, A. Semerci and N. Cobb, *Forest Ecol. Manage.*, 2010, **259**, 660–684.
3. K. R. Samabaraju, A. L. Carroll, J. Zhu, K. Stahl, R. D. Morre and B. H. Aukema, *Ecography*, 2012, **35**, 211–223.
4. A. L. Westerling, H. G. Hidalgo, D. R. Cayan and T. W. Swetnam, *Science*, 2006, **313**, 940–943.
5. D. McKenzie, D. W. Peterson, D. L. Peterson and P. E. Thornton, *J. Biogeogr.*, 2003, **30**, 1093–1108.
6. P. Mote, A. M. Hamlet, M. P. Clark and D. P. Lettenmaier, *Am. Meteor. Soc.*, 2005, **86**, 1–39.
7. C. M. Brasier, *Ann. Sci. For.*, 1996, **53**, 347–358.
8. H. H. Rogers, J. F. Thomas and G. E. Bingham, *Science*, 1983, **220**, 428–429.
9. C. Rathgeber, A. Nicault, J. Guiot, T. Keller, F. Guibal and P. Roche, *Global Planetary Change*, 2000, **26**, 405–421.

10. M. A. Drummond and T. R. Loveland, *BioScience*, 2010, **60**, 286–298.
11. J. Steinkamp and W. Peterman, http://app.databasin.org/app/pages/datasetPage.jsp?id=b2947eeae2e5488a86eacf0fcd4df7a4, last accessed 10.05.2012.
12. W. R. Anderegg, J. A. Berry, D. D. Smith, J. S. Sperry, L. D. Anderegg and C. B. Field, *Proc. Natl. Acad. Sci. U. S. A.*, 2012, **109**, 233–237.
13. K. A. Cherkauer and D. P. Lettenmaier, *J. Geophys Res.*, 1999, **104**, 19599–19610
14. Y. Iwata, M. Hayashi and T. Hirota, *Vadose Zone*, 2008, **7**, 79–86.
15. USGCRP, *Global Climate Impacts on the United States*, 2009; http://www.globalchange.gov/publications/reports/scientific-assessments/us-impacts, last accessed 10.05.2012.
16. S. Solomon, in *Climate Change 2007: The Physical Science Basis, Contribution of Working Group I to the Fourth Assessment Report of the Intergovernmental Panel on Climate Change*, ed. S. Solomon, D. Qin, M. Manning, Z. Chen, M. Marquis, K. B. Averyt, M. Tignor and H. L. Miller, Cambridge University Press, Cambridge, United Kingdom and New York, NY, USA, 2007.
17. S. Raper and R. J. Braithwaite, *Nature*, 2006, **439**, 311–313.
18. T. Kitzberger, T. W. Swetnam and T. T. Veblen, *Global Ecol. Biogeogr.*, 2001, **10**, 315–326.
19. Intergovernmental Panel on Climate Change (IPCC), *Fourth Assessment Report of the Intergovernmental Panel on Climate Change*, 2007, Cambridge Univ. Press, Cambridge.
20. P. Ciais, M. Reichstein, N. Viovy, A. Granier, J. Ogée, V. Allard, M. Aubinet, N. Buchmann, C. Bernhofer, A. Carrara, F. Chevallier, N. De Noblet, A. D. Friend, P. Friedlingstein, T. Grünwald, B. Heinesch, P. Keronen, A. Knohl, G. Krinner, D. Loustau, G. Manca, G. Matteucci, F. Miglietta, J. M. Ourcival, D. Papale, K. Pilegaard, S. Rambal, G. Seufert, J. F. Soussana, M. J. Sanz, E. D. Schulze, T. Vesala and R. Valentini, *Nature*, 2005, **437**, 529–533.
21. FAO, *Forestry Paper 147, Global Forest Resource Assessment 2005, Main Report*, 2006, pp. 11–36.
22. H. Hoogeveen, *Climate Change*, 2007, **2**, 36–46.
23. J. E. Aukema, D.G. McCullough, B. Von Holle, A.M. Liebold, K. Britton and S. J. Frankel, *BioScience*, 2010, **60**, 886–897.
24. A. M. Liebold, W. L. M. MacDonald, D. Bergdahl and V. C. Mastro, *Forest Sci.*, 1995, **41**, 1–49.
25. D. S. Wilcove, D. Rothstein, J. Bubow, A. Phillips and E. Losos, *BioScience*, 1998, **48**, 607–615.
26. D. Simberloff, *Nature and Human Society*, National Academy Press, Washington DC, USA, 2000.
27. E. A. Allen and L. M. Humble, *Can. J. Plant Pathol.*, 2002, **24**, 103–110.
28. J. P. Shepard, M. J. Mitchell, T. J. Scott, Y. M. Zhang and D. J. Raynal, *Water, Air Soil Pollut.*, 1989, **48**, 225–238.

29. K. C. Weathers, M. L. Cadenasso and S. T. A. Pickett, *Conserv. Biol.*, 2001, **15**, 1506–1514.
30. C. T. Driscoll, *BioScience*, 2001, **51**, 180–198.
31. K. C. Weathers and A. Ponette-González, in *Forest Hydrology and Biogeochemistry: Synthesis of Past Research and Future Directions*, ed. D. F. Levia, D.E. Carlyle-Moses and T. Tanaka, Ecological Studies Series, No. 216, Springer-Verlag, Heidelberg, Germany, 2011.
32. R. A. Skeffington and E. Wilson, *Environ. Pollut.*, 1988, **54**,159–184.
33. J. D. Aber, C. L. Goodale, S. V. Ollinger, M. L. Smith, A. H. Magill, M. E. Martin, R. A. Hallett and J. L. Stoddard, *BioScience*, 2003, **53**, 375–389.
34. S. E. Page, R. A. J. Wuest, D. Weiss, J. O. Riley, W. Shotyk and S. H. Limin, *J. Quat. Sci.*, 2004, **19**, 625–635.
35. J. Jaenicke, J. O. Riley, C. Mott, P. Kimman and F. Siegert, *Geoderma*, 2008, **147**, 151–158.
36. J. P. Andriesse, *FAO UN Soils Bulletin*,1988, **59**, 178.
37. R. A. Chimner and K. C. Ewel, *Wetlands Ecol. Manage.*, 2005, **13**, 671–684.
38. M. R. C. Posa, L. S. Wijedasa and R. T. Corlett, *BioScience*, 2011, **61**, 49–57.
39. A. Langner, J. Miettinen and F. Siegert, *Global Change Biol.*, 2007, **13**, 2329–2340.
40. A. Langner, F. Siegert, *Global Change Biol.*, 2009, **15**, 48–62.
41. M. R. Turetsky, W. F. Donahue and B. W. Benscoter, *Nature Commun.*, 2, 514.
42. S. E. Page, A. Hoscilo, A. Langner, K. Tansey, F. Siegert, S. Limin and J. Riley, in *Tropical Fire Ecology: Climate Change, Land Use and Ecosystem Dynamics*, ed. M. A. Cochrane, Springer-Praxis, Berlin, 2009, pp. 263–287.
43. U. Ballhorn, F. Siegert, M. Mason and S. Limin, *Proc. Natl. Acad. Sci. U. S. A.*, 2009, **106**, 21213–21218.
44. G. R. van der Werf, D. C. Morton, R. S. DeFries, J. G. J. Oliver, P. S. Kasibhatla, R. B. Jackson, G. J. Collatz and J. T. Randerson, *Nature Geosci.*, 2009, **2**, 737–739.
45. S. E. Page, F. Siegert, J. O. Rieley, H. -D. V. Boehm, A. Jaya and S. Limin, *Nature*, 2002, **420**, 61–65.
46. R. A. Betts, P. M. Cox, M. Collins, P. P. Harris, C. Huntingford and C. D. Jones, *Theor. Appl. Climatol.*, 2004, **78**, 157–175.
47. D. C. Nepstad, C. M. Stickler, B. S. Filho and F. Merry, *Philos. Trans. R. Soc. London, Ser. B*, 2008, **363**, 1737–1746.
48. B. S. Soares-Filho, D. C. Nepstad, L. M. Curran, G. C. Cerqueira, R. A. Garcia, C. A. Ramos, E. Voll, A. McDonald, P. LeFebvre and P. Schlesinger, *Nature*, 2006, **440**, 520–523.
49. J. G. Canadell, C. Le Quéré, M. R. Raupach, C. B. Field, E. T. Buitenhuis, P. Ciais, T. J. Conway, N. P. Gillett, R. A. Houghon and G. Marland, *Proc. Natl. Acad. Sci. U. S. A.*, 2007, **104**, 18866–18870
50. C. A. Nobre, P. J. Sellers and J. Shukla, *J. Climate*, 1991, **4**, 957–988.

51. P. M. Cox, R. A. Betts, C. D. Jones, S. A. Spall and I. J. Totterdell, *Nature*, 2000, **408**, 184–187.
52. P. M. Cox, R. A. Betts, M. Collins, P. P. Harris, C. Huntingford and C. D. Jones, *Theor. Appl. Climatol.*, 2004, **78**, 137–156.
53. A. Botta and J. A. Foley, *Global Biogoechem. Cycles*, 2002, **16**, 1070.
54. M. D. Oyama and C. A. Nobre, *Geophys Res Lett.*, 2003, **30**, 2199.
55. Y. Malhi, L. E. O. C. Aragão, D. Galbraith, C. Huntingford, R. Fisher, P. Zelazowski, S. Sitch, C. McSweeney and P. Meir, *Proc. Natl. Acad. Sci. U. S. A.*, 2009, **106**, 20610–20615.
56. R. Betts, M. Sanderson and S. Woodward, *Philos. Trans. R. Soc. London, Ser. B*, 2008, **363**, 1873–1880.
57. D. C. Nepstad, C. M. Stickler, B. S. Soares-Filho and F. Merry, *Philos. Trans. R. Soc. London, Ser. B*, 2008, **363**, 1737–1746.
58. S. Woodward, D. Roberts and R. Betts, *Geophys. Res. Lett.*, 2005, **32**, L18810.
59. J. Lean and P. R. Rowntree, *J. Climate*, 1997, **10**, 1216–1235.
60. D. C. Nepstad, C. R. De Carvalho, E. A. Davidson, P. H. Jipp, P. A. Le Febvre, G. H. Negreiros, E. S. Da Silva, T. A. Stone, S. E. Trumbore and S. Vieira, *Nature*, 1994, **372**, 666–669.
61. D. C. Nepstad, P. Lefebvre, U. Lopes da Silva, J. Tomasella, P. Schlesinger, L. Solórzano, P. Moutinho, D. Ray and J. Guerreira Benito, *Global Change Biol.*, 2004, **10**, 704–717.
62. D. C. Nepstad, I. M. Tohver, D. Ray, P. Moutinho and G. Cardinot, *Ecology*, 2007, **88**, 2259–2269.
63. M. G. Hodnett, L. Pimental da Silva, H. R. da Rocha and R. Cruz Senna, *J. Hydrol.*, 1995, **170**, 233–254.
64. R. S. Oliveira, R. E. Dawson, S. S. O. Burgess and D. C. Nepstad, *Oecologia*, 2005, **145**, 354–363.
65. P. M. Brando, D. C. Nepstad, E. A. Davidson, S. E. Trumbore, D. Ray and P. Camargo, *Philos. Trans. R. Soc. London, Ser. B*, 2008, **363**, 1839–1848.
66. G. B. Williamson, W. F. Laurence, A. A. Oliveira, P. Delamônica, G. Gascon, T. E. Lovejoy and L. Pohl, *Conserv. Biol.*, 2000, **14**, 1538–1542.
67. D. Ray, D. C. Nepstad and P. Montinho, *Ecol. Appl.*, 2005, **15**, 1664–1678.
68. L. R. Iverson and A. M. Prasad, *Ecosystems*, 2001, **4**, 186–199.
69. USDA, Forest Inventory Analysis, 2009; http://fia.fs.fed.us/library/bus-org-documents/, last accessed 10.05.2012.
70. N. Coops, R. H. Waring and B. E. Law, *Ecol. Modelling*, 2005, **183**, 107–124.
71. N. Coops and R. H. Waring, *Climate Change*, 2010, **105**, 313–328.
72. N. Coops, R. H. Waring, C. Beier, R. R. Jauvin and T. Wang, *Appl. Vegetat. Sci.*, 2011, **14**, 402–414.
73. A. Fajardo, E. J. B. McIntire, *J. Ecol.*, 2012.
74. J. Lenoir, J. C. Gégout, P. A. Marquet, P. de Ruffray and H. Brise, *Science*, 2008, **320**, 1768–1771.
75. F. Archaux, *Ibis*, 2004, **146**, 138.
76. S. Popy, L. Bordignon and R. Prodon, *J. Biogeogr.*, 2010, **37**, 57.

77. S. M. Crimmins, S. Z. Dobrowski, J. A. Greenberg, J. T. Abatzoglou and A. R. Mynsberge, *Science*, 2011, **331**, 324–327.
78. C. I. Millar, L. J. Graumlich, D. L. Delany, R. D. Westfall and J. K. King, *Arct. Alp. Res.*, 2004, **36**, 181.
79. A. J. Eckert and M. L. Eckert, *Madrono*, 2007, **54**, 117.
80. A. Kattenberg A. F. Giorgi, H. Grassl, G. A. Meehl, J. F. B. Mitchell, R.J. Stouffer, T. Tokioka, A. J. Weaver and T. M. L. Wigley, in *The Science of Climate Change: Contribution of Working Group I to the Second Assessment of the IPCC*, 1996, Cambridge University Press, UK, pp. 285–357.
81. M. C. Serreze, J. E. Walsh, F. S. Chapin, T. Osterkamp, M. Dyurgerov, V. Romanovsky, W. C. Oechel, J. Morison, T. Zhang and R. G. Barry, *Climate Change.*, 2000, **46**, 159–207.
82. L. A. Viercek and C. T. Dyrness, in *Forest Ecosystems in the Alaskan Taiga: A Synthesis of Structure and Function*, ed. K. Van Cleve, F. S. Chapin, III, P. W. Flanagan, L. A. Vierick and C. T. Dyrness, Springer-Verlag, New York, 1986, pp. 22–43.
83. C. Potter, *Ecol. Modelling*, 2004, **175**, 1–24.
84. R. V. Densmore, *Am. J. Bot.*, 1997, **84**, 274–278.
85. D. V. D'Amore and P. E. Hennon, *Global Change Biol.*, 2006, **12**, 524–545.
86. P. E. Hennon, C. G. Shaw III and E. M. Hansen, *Plant Disease*, 1990, **74**, 267–273.
87. C. G. Shaw III, A. Eglitis, T. H. Laurent and P. E. Hennon, *Plant Disease*, 1985, **69**, 13–17.
88. P. E. Hennon, G. B. Newcomb, C. G. Shaw III and E. M. Hansen, *Plant Disease*, 1986, **70**, 352.
89. E. M. Hansen, P. B. Hamm, C. G. Shaw III and P. E. Hennon, *Trans. Br. Mycolog. Soc.*, 1988, **91**, 379–384.
90. P. B. Hamm, E. M. Hansen, P. E. Hennon and C. G. Shaw III, *Trans. Br. Mycolog. Soc.*, 1988, **91**, 385–388.
91. P. E. Hennon, *Mycologia*,1990, **82**, 59–66.
92. P. E. Hennon, E. M. Hansen and C. G. Shaw III, *Northwest Sci.*, 1990, **64**, 45–54.
93. P. E. Hennon and M. J. McWilliams, *Can. J. Forest Res.*, 1999, **29**, 1985–1988.
94. J. A. Antos and D. B. Zobel, *Can. J. Bot.*, 1986, **64**, 1898–1909.
95. W. L. Chapman and J. E. Walsh, *Bull. Am. Meteorol. Soc.*, 1993, **74**, 33–47.
96. R. J. Hebda, *Geogr. Phys. Quat.*, 1995, **49**, 55–79.
97. FAO, *State of the World's Forests*, Rome, 2011; http://www.fao.org/docrep/013/i2000e/i2000e00.htm, last accessed 10.05.2012.
98. J. A. Jones, G. L. Achterman, L. A. Augustine, I. F. Creed, P. F. Folliott, L. MacDonald and B. C. Wemple, *Hydrol Process*, 2009, **23**, 2699–2704.
99. USGCRP; http://www.globalchange.gov/component/content/article/52-reports-and-assessments/345-looking-at-americas-climate, last accessed 05.04.2012.
100. J. Tang and Q. Zhuang, *Climatic Change*, 2011, **107**, 481–510.

101. J. S. Kimball, K. C. McDonald and M. Zhao, *Earth Interact.*, 2006, **10**, 1–22.
102. G. A. Meehl and T. F. Stocker, in *Climate Change 2007: the Physical Science Basis*, ed. S. Solomon, D. Qin, M. Manning, Z. Chen, M. Marquis, K. B. Averyt, M. Tignor and H. L. Miller, Cambridge University Press, Cambridge, 2007.
103. G. Saramiento, in *High Altitude Tropical Biogeography*, ed. F. Vuilleumier and M. Monasterio, Oxford University Press, Oxford, 1986, pp. 11–45.
104. K. S. H. Peh, M. C. K. Soh, N. S. Sodhi, W. F. Laurence, D. J. Ong and R. Clements, *BioScience*, 2011, **61**, 27–38.
105. H. J. Boehmer, in *Coping with Global Environmental Change, Distasters and Security*, ed. H. G. Brauch, U. O. Spring, C. Mesjasz, J. Grin, P. Kameri-Mbote, B. Chourou, P. Dunay and J. Birkman, Springer-Verlag, Berlin-Heidelberg-New York, 2011, p. 790–802.
106. N. S. Sodhi and B. W. Brook, *Southeast Asia Biodiversity in Crisis*, Cambridge University Press, Cambridge, UK, 2006.
107. L. Ong, MSc thesis, National University of Singapore, 2000.
108. P. Bubb, I. May, L. Miles and J. Sayer, *Cloud Forest Agenda*, UN Environment Programme World Conservation Monitoring Centre, 2004.
109. T. K. Waggener, in *Asia-Pacific Forestry Sector Outlook Study – Working Paper Series*, FAO, 1997.
110. C. Doumenge D. Gilmour, M. R. Perez and J. Blockhus, in *Tropical Montane Cloud Forests*, ed. L. S. Hamilton, J. S. Juvik and F. N. Scatena, 1995, pp. 24–37.
111. K. W. Leong and N. W. Chan, in *Cameron Highlands*, ed. N. W. Chan, 2006, pp. 121–132.
112. L. A. Bruijnzeel and E. J. Veneklaas, *Ecology*, 1998, **79**, 3–9.
113. USGCRP, *Global Climate Impacts on the United States*, 2009; http://www.globalchange.gov/publications/reports/scientific-assessments/us-impacts, last accessed 10.05.2012.
114. T. Hudiburg, B. Law, D. P. Turner, J. Campbell, J. Donato and M. Duane, *Ecol. Appl.*, 2009, **19**, 163–180.
115. K. Lajtha and F. J. Barnes, *Tree Physiol.*, 1991, **9**, 59–67.
116. C. Potter, *Ecol. Modelling*, 2004, **175**, 1–24.
117. G. P. Asner and S. Archer, in *Livestock in a Changing Landscape: Drivers, Consequences and Responses*, ed. H. Steinfeld, H. Mooney, F. Schneider, L. E. Neville, Washington, DC, Island Press, 2010, pp. 69–82.
118. K. A. Hibbard, S. Archer, D. S. Schimel and D. W. Valentine, *Ecology*, 2001, **82**, 1999–2011.
119. R. F. Hughes, G.P. Asner, C. A. Wessman, C. McMurtry, J. Nelson and R. J. Ansley, *Global Change Biol.*, 2006, **12**, 1733–1747.
120. A. K. Knapp, J. M. Briggs, S. L. Collins, S. R. Archer, M. S. Bret-Harte, B. E. Ewers, D. P. Peters, D. R. Young, G. R. Shaver, E. Pendall and M. B. Cleary, *Global Change Biol.*, 2008, **14**, 615–623.

121. N. N. Barger, S. R. Archer, J. L. Campbell, C. H. Huang, J. Morton and A. K. Knapp, *J. Geophys. Res.*, 2011, **116**, 1–17.
122. W. H. Schlesinger, *Biogeochemistry: An Analysis of Gobal Ghange*, Academic Press, New York, 2nd edn, 1997.
123. T. B. Bragg and L. C. Hulbert, *J. Range Manage*, 1976, **29**, 19–24.
124. S. Archer, C. J. Scifres, C. Bassham and R. Maggio, *Ecol. Monogr.*, 1988, **58**, 111–127.
125. D. R. Lane, D. P. Coffin and W. K. Lauenroth, *J. Veg. Sci.*, 1998, **9**, 239–250.
126. R. J. Ansley and M. J. Castellano, *Restorat. Ecol.*, 2006, **14**, 420–428.
127. J. M. Briggs, G. A. Hoch and L. C. Johnson, *Ecosystems*, 2002, **5**, 578.
128. S. C. Goslee, K. M. Harstad, D. P. C. Peters, A. Rango and W. H. Schelsinger, *Arid Environ.*, 2003, **54**, 755–767.
129. R. B. Jackson, H. J. Schenk, E. G. Jobbágy, J. Canadell, G. D. Colello, R. E. Dickinson, C. B. Field, P. Friedlingstein, M. Heimann, K. Hibbard, D. W. Kicklighter, A. Kleidon, R. P. Neilson, W. J. Parton, O. E. Sala and M. T. Sykes, *Ecol. Appl.*, 2000, **10**, 470–483.
130. R. B. Jackson, J. L. Banner, E. G. Jobbágy, W. T. Pockman and D. H. Wall, *Nature*, 2002, **418**, 623–626.
131. C. A. Wessman, S. Archer, L. C. Johnson and G. P. Asner, in *Land Change Science: Observing, Monitoring and Understanding Trajectories of Change on the Earth*, ed. G. Gutman, A. C. Janetos, C.O. Justice, E. F. Moran, J. F. Mustard, R. R. Rindfuss, D. Skole, B. L. Turner II and M. A. Cochrane, Kluwer Acad., Dordrecht, Netherlands, 2004, pp. 185–208.
132. S. Archer, in *Ecological Implications of Livestock Herbivory in the West*, ed. M. Vavra, W. A. Laycock and R. D. Pieper, Society for Range Management, Denver, CO, 1994, pp. 13–68
133. S. Archer, D. S. Schimel and E. A. Holland, *Climate Change*, 1995, **29**, 91–99.
134. O. W. Van Auken, *Annu. Rev. Ecol. Syst.*, 2000, **31**, 197–215.
135. E. G. Jobbàgy and R. B. Jackson, *Ecol. Appl.*, 2000, **10**, 423–436.
136. T. W. Boutton, J. D. Liao, T. R. Filley, S. R. Archer and R. Follett, in *Soil Carbon Sequestration and the Greenhouse Effect*, ed. R. Lal and R. Follett, 2008, pp. 181–205.
137. Y. Bai, T. Colberg, J. T. Romo, B. McConkey, D. Pennock and R. Farrell, *Agric. Ecosyst. Environ.*, 2009, **134**, 269–276.
138. J. Canadell, R. B. Jackson, J. R. Ehleringer, H. A. Mooney, O. E. Sala and E. D. Schulze, *Oecologia*, 1996, **108**, 583–595.
139. R. B. Jackson, J. Canadell, J. R. Ehleringer, H. A Mooney, O. E. Sala and E. D. Schulze, *Oecologia*, 1996, **108**, 389–411.
140. T. W. Boutton, S. R. Archer, A. J. Midwood, S. F. Zitzer and R. Bol, *Geoderma*, 1998, **82**, 5–41.
141. L. A. Biderman and T. W. Boutton, *Soil Biol. Biochem.*, 2009, **41**, 1943–1950.
142. A. J. Franzluebbers and M. A. Arshad, *Can. J. Soil Sci.*, 1996, **76**, 387–393.
143. A. J. Franzluebbers and M. A. Arshad, *Soil Sci. Soc. Am. J.*, 1997, **61**, 1382–1386.

144. R. A. Gill and I. C. Burke, *Oecologia*, 1999, **121**, 551–563.
145. J. D. Liao, T. W. Boutton and J. D. Jastrow, *Soil Biol. Biochem.*, 2006, **38**, 3184–3196.
146. C. J. Field, J. Merino and H. A. Mooney, *Oecologia*, 1983, **60**, 384–389.
147. K. Lajtha and W. G. Whitford, *Oecologia*, 1989, **80**, 341–348.
148. H. G. Miller and J. D. Miller, *J. Appl. Ecol.*, 1976, 249–256.
149. H. Brix, *Can. J. For. Res.*, 1981, **11**, 502–511.
150. M. R. Keyes, C. C. Grier, *Can. J. For. Res.*, 1981, **11**, 599–605.
151. S. Linder and D. A. Rook, in *Nutrition of Plantation Forests*, ed. G. D. Bowen, E. K. S. Nambiar, Academic Press, London, 1984, pp. 211–263.
152. K. J. Nadelhoffer, J. D. Aber and J. M. Melillo, *Ecology*, 1985, **66**, 1377–1390.
153. S. Linder, in *Potentials and Limitations of Ecosystem Analysis*, ed. E. D. Schulze and H. Z. Wolfer, Springer-Verlag, Berlin, 1987, pp. 180–202.
154. B. J. Myers, *Tree Physiol.*, 1988, **4**, 315–323.
155. D. Sanantonio, in *Biomass Production by Fast-Growing Trees*, ed. J. S. Pereira and J. J. Landsberg, Kluwer Academic Publishers, The Netherlands, 1989, pp. 57–72.
156. R. J. Raison and R. Stottlemyer, *Tree Physiology*, 1991, **9**, 209–225.
157. J. L. Charley in *Wildland Shrubs –Their Biology and Utilization*, ed. C. M. McKell, J. P. Blaisdell and J. R. Goodwin, Intermountain Forest and Range Research Station, US Department of Agriculture, Ogden, Utah, 1972, pp. 182–203.
158. J. A. Ludwig in *Larrea Derie el Desierto*, ed. E. C. Lopez, T. J. Mabry and S. F. Tavizopp, Centro de Investigacion en Quimica Aplicada, Saltillo, Coahila, Mexico, 1979, pp. 139–150.
159. J. F. Reynolds and G. L. Cunningham, *Oecologia*, 1981, **51**, 357–363.
160. A. Ayanaba, S. B. Tuckwell and D. S. Jenkinson, *Soil Biol. Biochem.*, 1976, **8**, 519–525.
161. J. M. Shurer, *Soil Biol. Biochem.*, 1985, **17**, 611–618.
162. D. S. Powlson and P. C. Prookes, *Soil Biol. Biochem.*, 1987, **19**, 159–164.
163. S. Perakis and E. R. Sinkhorn, *Ecology*, 2011, **92**, 1481–1491.
164. L. M. Christensen, G. M. Lovett, K. Weathers and M. Arthur, *Soil Sci. Soc. Am. J.*, 2009, **73**, 638–646.
165. P. N. Beets and H. A. I. Madgwick, *N. Z. J. For. Sci.*, 1988, **18**, 43–64.
166. S. K. Chapman, J. A. Langley, S. C. Hart and G. W. Koch, *New Phytologist*, 2006, **169**, 27–34.
167. G. E. Likens, F. H. Bormann, R. S. Pierce, J. S. Eaton and N. M. Johnson, *Biogeochemistry of a Forested Ecosystem*, Springer-Verlag, New York-Heidelberg-Berlin, 1977, p. 146.
168. P. M. Vitousek and J. M. Melillo, *For. Sci.*, 1979, **25**, 605–619.
169. P. M. Vitousek, *Ecology*, 1984, **66**, 1360–1376.
170. R. J. Raison, P. K. Khanna and P. V. Woods, *Can. J. For. Res.*, 1985, **15**, 657–664.
171. A. M. O'Connell, *For. Ecol. Manage.*, 1989, **26**, 95–111.

172. R. J. Raison, P. K. Khanna and P. V. Woods, *Can. J. For. Res.*, 1985, **15**, 132–140.
173. N. W. Foster, J. A. Nicolson and P. W. Hazlett, *J. Environ. Qual.*, 1989, **18**, 238–244.
174. R. Stottlemyer and D. Rutkowski. *Water Resour. Res.*, 1990, **26**, 721–737.
175. R. Lea, W. C. Tierson, D. H. Bickelhaupt and A. L. Leaf, *Plant Soil*, 1980, **54**, 419–439.
176. H. L. Mitchell, *Black Forest Bull.*, 1939, 11.
177. C. M. Rascher, C. T. Driscoll and N. E. Peters, *Biogeochemistry*, 1987, **3**, 209–224.
178. S. Bernal, L. O. Hedin, G. E. Likens, S. Gerber and D. C. Buso, *Proc. Natl. Acad. Sci. U. S. A.*, 2012, **109**, 3406–3411.
179. J. Kaiser, *Science*, 2004, **304**, p. 1616.
180. J. L. Smith, J. J. Halvorson and R. I. Papendick, *Soil Sci. Am. J.*, 1994, **57**, 743–749.
181. L. R. Oldeman, in *ISRIC Bi-Annual Report*, Netherlands, 1992, pp. 19–36.
182. R. Lal, *Science*, 2004, **304**, 1623.
183. J. Gaillardet, B. Dupre, P. Louvat and C. J. Allegre, *Chem. Geol.*, 1999, **159**, 3–30.
184. P. A. Raymond and J. J. Cole, *Science*, 2003, **301**, 88–91.
185. G. Bayon, B. Dannielou, J. Etoubleau, E. Ponzevera, S. Toucanne and S. Bernell, *Science*, 2012, **335**, 1219–1222.
186. T. N. Huffman, *Annu. Rev. Anthropol.*, 1982, **11**, 133.
187. J. Vansina, *J. Afr. Hist.*, 1984, **25**, 129.
188. D. W. Phillipson, *African Archaeology*, Cambridge University Press, Cambridge, 1993.
189. C. J. Holden, Proc. R. Soc. London, Ser. B, 2002, **269**, 793.
190. J. Diamond and P. Bellwood, *Science*, 2003, **300**, 597.
191. K. J. Willis, L. Gillson and T. M. Brncic, *Science*, 2004, **304**, 402.
192. A. Vincens, D. Schwartz, H. Elenga, I. Reynaud-Farrera, A. Alexandre, J. Bertaux, A. Mariotti, L. Martin, J. -D. Meunier, F. Nguetsop, M. Servant, S. Servant-Vildary and D. Wirrmann, *J. Biogeogr.*, 1999, **26**, 879.
193. J. A. Maley, *Inst. Dev. Studies Bull.*, 2002, **33**, 13.
194. F. J. Stevenson, *Cycles of Soil: Carbon, Nitrogen, Phosphorus, Sulfur, Micronutrients*, John Wiley and Sons, New York, 1999.
195. F. Khormali, M. Ajami, S. Ayoubi, Ch. Srinivasarao and S. P. Wani, *Agric. Ecosyst. Environ.*, 2009, **134**, 178–189.
196. T. G. Vagen, *Geoderma*, 2006, **131**, 190–200.
197. M. Lemenih and D. Teketay, *Agric. Ecosyst. Environ.*, 2005, **105**, 373–386.
198. J. Six, K. Paustian, E. T. Elliott and C. Combrink, *Soil Sci. Soc. Am. J.*, 2000, **64**, 681–689.
199. V. Rasiah, S. K. Florentine, B. L. Williams and M. E. Westbrooke, *Geoderma*, 2004, **120**, 35–45.
200. K. Zhang, S. Li, W. Peng and B. Yu, *Soil Till Res.*, 2004, **76**, 157–165.
201. R. B. Jackson and J. S. Baker, *BioScience*, 2010, 2010, **60**, 698–707.

202. C. Conte, *BioScience*, 2010, **60**, 309–313.
203. L. Rival, in *Advances in Historical Ecology*, ed. W. Balee, Columbia University Press, New York, 1998, pp. 232–250.
204. E. Krieke, in *History and African Environments*, ed. W. Beinert and J. McGregor, Currey, Oxford, 2003, pp. 27–42.
205. S. Feierman, *The Shambaa Kingdom*, University of Wisconsin Press, Madison, 1974.
206. J. E. G. Sutton, *Azania*, 1989, **24**, 98–112.
207. P. Schmidt, in *Historical Ecology Cultural Knowledge and Changing Landscape*, ed. C. Crumley, School of American Research Press, Santa Fe, 1994, pp. 232–250.
208. J. Vansina, *J. Afr. Hist.*, 1995, **36**, 173–195.
209. C. Ehret, *An African Classical Age: Eastern and Southern Africa in World History, 1000 BC to AD 400*, University of Virginia, Charlottesville, 2001.
210. A. C. Hamilton, *Environmental History of East Africa: A Study of the Quaternary*, Academic Press, New York, 1982.
211. O. Warburg, *Die Kulturpflanzen Usambaras, Mitteilungen aus den Deutschen Schutzgebeiten*, 7, 1894, pp. 131–199.
212. B. J. Lejju, P. Robertshaw and D. Taylor, *J. Arch. Sci.*, 2006, **33**, 102–113.

Plant Nutrients

DAVID A. C. MANNING

School of Civil Engineering and Geosciences, Newcastle University, Newcastle upon Tyne, NE1 7RU, UK
E-mail: david.manning@newcastle.ac.uk

ABSTRACT

All plants require nitrogen (N), phosphorus (P) and potassium (K) as major nutrients, from a range of possible sources including artificial fertilisers. Both P and K (and almost all minor and trace nutrients) are derived from mined sources: P from phosphate rock and K from potassium salts such as sylvite (KCl). The reserves of P and K are equivalent to up to 400 years production at current rates, and resources have life expectancies of 1800 and 7500 years, respectively. Prices of K fertilisers are currently very high (US$ 500 per tonne), in part reflecting the cost of mining. However, nutrient audit studies for the end of the 20th Century show that although N is approximately in balance, 30% more P and twice as much K needs to be mined to compensate for that removed by crops. With growing global populations, world production of K needs to triple by 2050 to feed the expected population, whereas P production needs to increase by 70%. In these circumstances, there is a pressing need to broaden the range of available sources of K, especially for farmers who cannot afford conventional fertilisers. Candidates include the silicate minerals, such as feldspars, feldspathoids and micas. Mineral dissolution rates show that feldspathoids dissolve 10^5 –10^7 times more rapidly than feldspars, and in contrast micas release K by

Issues in Environmental Science and Technology No. 35
Soils and Food Security
Edited by R.E. Hester and R.M. Harrison

Published by the Royal Society of Chemistry, www.rsc.org

cation exchange. Given the current high price of K, it is appropriate to consider widely available silicate minerals as an alternative source.

1 Introduction

This chapter focuses on the need for mineral fertilisers, recognising that these are supplied by the mining industry and ultimately come from finite (but very large) stocks of geological materials.

1.1 Geological Sources of Plant Nutrients

For healthy growth, plants require a diverse range of nutrients. Carbon dioxide is derived from the air, as a raw material for photosynthesis, and water is derived from surface or groundwater sources. The dominant fertiliser nutrients that are applied to farmed plants are nitrogen (N), phosphorus (P) and potassium (K), which make up the bulk of the global fertiliser industry output. Minor nutrients and trace elements are also vital, and can be applied artificially if there is evidence of deficiencies.

The fertiliser industry is dominated by the production of N, P and K fertilisers as bulk commodities. This is big business, worth of the order of US$ 70 billion for N (fixed ammonium),[1] US$ 20 billion for P,[2] and US$ 26 billion for K.[3] By comparison, the global petroleum industry is worth of the order of US$ 2000 billion, gas US$ 750 billion, and coal US$ 1000 billion.[4]

The production of N, P and K fertilisers depends on geological resources, but these differ in each case.[5] Nitrogen production is very closely related to that of natural gas, on two counts – gas is a source of raw materials and a source of the energy required to fuel the process. N fertiliser production uses the Haber process, which involves reaction between atmospheric N and hydrogen derived from methane, to produce ammonia and carbon dioxide, according to the following simplified reactions:

$$CH_4 + 2H_2O = CO_2 + 4H_2 \quad (1)$$

$$N_2 + 3H_2 = 2NH_3 \quad (2)$$

The CO_2 produced in this process is not necessarily wasted, as it can be combined with ammonium to produce urea, $CO(NH_2)_2$, a common fertiliser product. Nitrogen fertiliser production is energy-intensive, consuming 94% of the energy used globally in the manufacture of all fertilisers (K and P use 3% each).[5]

In contrast to N, both P and K fertilisers are derived from mined rocks, with varying amounts of processing. Phosphate fertilisers are ultimately derived uniquely from phosphate rock, which occurs widely across the globe.

Sedimentary phosphate rocks, such as those of North Africa and the Middle East, are the dominant sources, and igneous phosphate rocks are also mined. In all of these, the most important phosphate mineral is apatite, $Ca_5(PO_4)_3(OH)$; the OH site can contain carbonate (sedimentary sources) or fluoride (igneous sources). Chemical processing during manufacture varies from nil, when phosphate rock is simply sold as a crushed and ground material suitable for direct application, to chemical manufacturing processes that produce phosphoric acid. This is then used as a raw material both for the manufacture of fertilisers and other industrial chemical feedstocks and materials, such as detergents. When used directly as a fertiliser, without chemical processing, phosphate rock is a slow-release source of P, in which the availability of phosphate depends on the dissolution of apatite in the soil solution:

$$Ca_5(PO_4)_3(OH) + H^+ = 5Ca^{2+} + 3PO_4^{3-} + H_2O \tag{3}$$

Potassium fertilisers differ from chemically processed N and P fertilisers in that they are mined as a readily soluble salt. Geological deposits of potassium salts, generally termed 'potash', occur within rocks known as evaporites, produced by the evaporation of saline brines in very specific geological processes; these are mimicked in the artificial production of salts from natural brines, for example from the Dead Sea brines. Examples of potash minerals are given in Table 1. Once mined, typically they are either used directly, with limited processing, or processed to separate specific components. They may be blended with other fertiliser products to give a compound fertiliser.

Table 1 Potash minerals. (Note that it is conventional to express the potash content in terms of the equivalent amount of K_2O).

Mineral name	*Chemical formula*	*Potash content (%K_2O)*
Sylvite	KCl	63%
Carnallite	$KCl.MgCl_2.6H_2O$	19%
Langbeinite	$K_2SO_4.2MgSO_4$	23%
Kainite	$4KCl.4MgSO_4.11H_2O$	19%
Polyhalite	$K_2SO_4.2CaSO_4.MgSO_4.2H_2O$	16%

1.2 Minerals in Plants

Chemically, plants are complex systems. Their composition is dominated by water and carbon compounds, and they also contain inorganic components, commonly described as the mineral content of a plant, a crop or a food. This is the residue that is left behind as an ash following complete combustion of plant material.

Table 2 shows that the mineral content of many crops is typically quite low, less than 1% by weight for each element. In addition to the major nutrients shown in Table 2 (P and K), plant growth depends on the availability of minor nutrients (Ca, Mg, Si and S), and trace nutrients (including B, Fe, Mn, Zn, Se and others). All of these are ultimately derived from geological sources. Table 3 summarises the major mineral sources for these elements. In some cases, minor nutrients especially can be obtained from minerals in which the element concerned is a major component (for example, the double salts of Mg and K shown in Table 1, also carbonate minerals or gypsum). In other cases, element availability depends on the weathering of silicate minerals such as tourmaline (a complex borosilicate, which supplies B), or silicates containing Fe, Mg or Ca. Other elements such as Zn and Se may be derived by the weathering of rocks such as mudstones, in which the mineralogical source of the element may not be obvious, and in which elements such as Se might be associated with organic matter.

The availability of nutrients needs to be considered in the scale of crop production. With the removal of each crop from soil, mineral nutrients are also

Table 2 Mineral content of selected food and energy crops.[6,7]

Crop	*P*	*K*	*Ca*	*Mg*	*Si*
	mg $(100\ g)^{-1}$	mg $(100\ g)^{-1}$	mg $(100\ g)^{-1}$	mg $(100\ g)^{-1}$	mg $(100\ g)^{-1}$
Sugar cane	200	900	200	120	100
Wheat	100	2000	300	200	4000
Potato	37	360	5	17	
Leek	44	260	24	3	
Apple	11	120	4	5	
Banana	28	400	6	34	

Table 3 Mineral sources of minor and trace nutrients in plants.

		Mineral source
Minor nutrients		
Magnesium	Mg	Carnallite, langbeinite, kainite, polyhalite; dolomite ($CaMg(CO_3)_2$); weathering of Mg-Fe silicate minerals
Calcium	Ca	Lime ($Ca(OH)_2$) or limestone ($CaCO_3$); gypsum ($CaSO_4.2H_2O$); dolomite; weathering of Ca silicate minerals
Sulfur	S	Gypsum
Silicon	Si	Weathering of silicate minerals
Trace nutrients		
Boron	B	Weathering of tourmaline
Iron	Fe	Weathering of Fe-Mg silicate minerals
Manganese	Mn	Weathering of Fe-Mg silicate minerals
Zinc	Zn	Weathering of sulfide minerals
Selenium	Se	Weathering of certain mudstones

Figure 1 Moving sugar cane harvest from field to factory, Thailand. Each lorry carries around 10 T of cane, containing 0.1 T of K that will cost around US$ 50 to replace.

removed from the soil as 'offtake'. The principle of sustainable fertiliser use is to design application rates that compensate for offtake,[8] so that the soil is not mined of its nutrient content and is ready, nutritionally, to supply the next crop. In terms of demand, it is the major nutrients (N, P and K) that dominate world fertiliser markets, and minor and trace nutrients will not be considered further in this chapter.

As an example of the scale of removal of major nutrients from cropped land, Thailand produces 100 million tonnes of sugar cane per year,[9] representing the removal of 1 million tonnes of K from the soil. A lorry carrying 10 tonnes of sugar cane (Figure 1) is removing 100 kg of K from the soil, and this will cost around US$ 50 to replace. When considering fertiliser minerals and their use, it is important to understand their availability and price.

2 Availability of P and K Fertilisers: Supply, Demand and Price

World production of P and K fertilisers is summarised in Figure 2, which shows annual production and declared reserves. It is important to note that the amount of a commodity that is declared as a reserve is dictated not by geology, but by the reporting requirements of the legally binding codes adopted by the commodity traders – to protect investors. The term 'reserve' refers to material of known grade that is present in the ground in a known quantity and that can be mined using present technology to generate a profit.[10] In contrast, the term 'resource' refers to a material known to exist in the ground, that is suitable for further development (in terms of mineral exploration and deposit modelling) so that additional reserves can be defined.

Figure 2 shows a dramatic increase in the reserves of phosphate rock in 2010. This is because of reclassification of material following a change in the application of the reporting codes. Additionally, extremely large deposits of

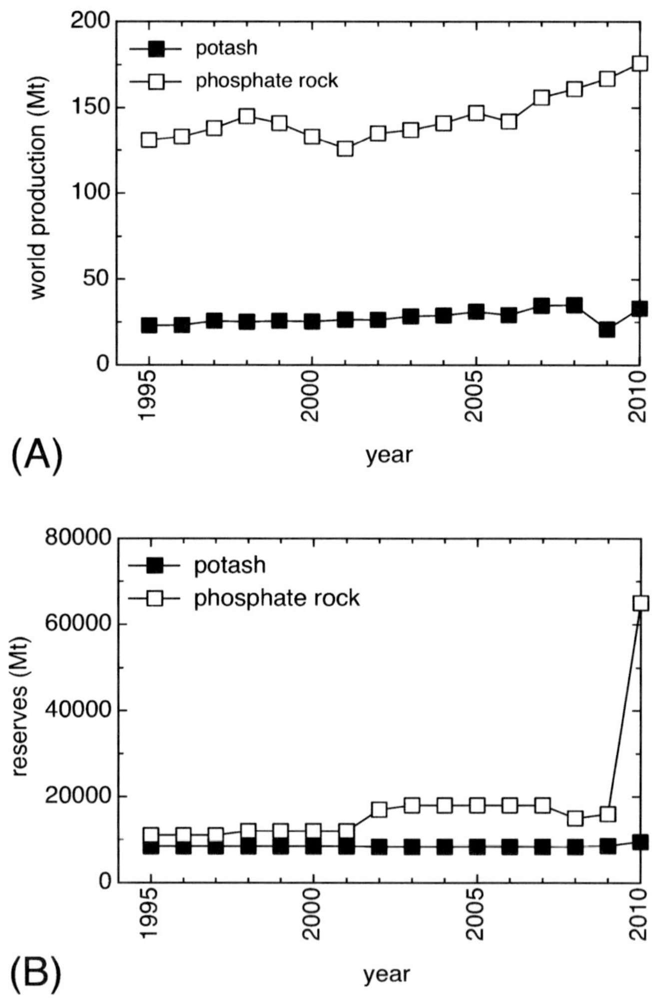

Figure 2 Annual rates of production (A) and declared reserves (B) for K and P fertiliser minerals.[2,3]

phosphate rock were reported from Iraq, adding significantly to the total known reserves.

Potash is mined at the rate of 25–30 million tonnes K_2O annually. Just 12 countries produce 99% of the world's potash, with over 30% of production derived from Canada, 50% from the USA and Canada combined, where over 90% of the world's reserves are located. There is no potash production of significance from Africa, south/south-east Asia or Australasia. In contrast, over 30 countries mine significant quantities of phosphate rock, in all continents. China is the dominant producer. According to the Food and Agriculture Organization of the United Nations (FAO), fertiliser production is sufficient to meet demand from world markets.[11]

At current rates of mining, declared reserves of K have a lifetime of between 200 and 400 years, and those of P have a lifetime of 400 years. In terms of their life expectancy, estimated resources of K are equivalent to 7500 years, and of P are equivalent to 1800 years, assuming production at current levels. Given these figures, it is important to be cautious when considering the concept of 'peak phosphorus'.[12] Also, waste-waters represent an excellent potential source of P that can be recovered for use as a fertiliser.[13]

The price of fertilisers has been highly variable in recent years. Figure 3 shows changes in the price of diammonium phosphate (DAP), urea and KCl fertilisers since 2000. The price of urea precisely tracks the price of Brent crude

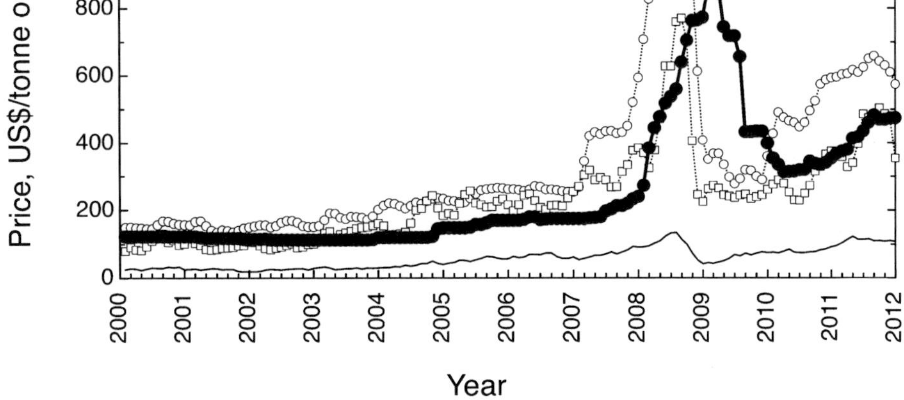

Figure 3 Variation in prices of phosphorus (diammonium phosphate; DAP), nitrogen (urea) and potassium (muriate of potash; MOP) fertilisers since 2000. (Source: plotted using data from the World Bank, April 2012, http://databank.worldbank.org).

oil (reflecting urea manufacture's high energy requirement); DAP does so to a lesser extent. Urea and DAP fertiliser prices peaked in 2008, reflecting changes in the price of petroleum. KCl peaked soon afterwards, reaching almost US$ 900 per tonne. Prices dropped in 2009, and have since started to rise again. At present, the price of KCl is about US$ 500 per tonne.

3 Nutrient Audits and Fertiliser Use Statistics – Evidence of Need

Farmers use fertilisers to replace the nutrients that crops have removed from the soil. After harvest, nutrients can be returned to the soil from crop residues; they are also supplied by manures and composts, often produced locally. Fertilisers are purchased on a market that has a limited number of suppliers focused in geographically few locations.

Audits of plant nutrient supply underpin the use of fertilisers, and give rise to the concept of nutrient balances. In the UK, recommended fertiliser application rates are published by the government,[8] and these tell farmers what applications are required for different crops so that soil fertility is maintained from year to year. GPS-controlled precision methods can be used to take into account natural variability in natural soil nutrient content within a single field, to ensure that just the right amount of fertiliser is applied.

On a wider scale, nutrient audits for P[14,15] and P, K, and N[16] have been carried out globally and for individual countries. These assess the amount of nutrient removed by offtake (removal of a crop from the field) and compare this with the amounts of nutrient that are supplied by different inputs. An example is shown in Figure 4. This shows that, on an assessment of the entire continent of Africa,[17] total nitrogen inputs match outputs, and so N is in balance. Nitrogen fertiliser inputs supply less than 30% of total inputs. Similarly, P is also in balance or added to excess; in this case, fertilisers make up 50% or more of inputs. The situation with K differs; total inputs correspond to 40% of outputs, and fertiliser use corresponds to no more than 20% of inputs. Thus, on a continental basis, there is removal of K from African soils that is not replenished by fertiliser inputs. This observation is supported by data published by the FAO.[11] Of the 57 African countries considered, 47 consume no potash fertiliser, and the total imports of K fertiliser (there is no significant indigenous production) amount to 450 000 tonnes, 1.5% of world production in support of almost 15% of the world's population.

On a global basis for the period 1960–1996, it has been shown[16] that the net annual rates of removal of N and P from soil progressively increase from 8 to 12 kg N ha^{-1} and from 2 to 4 kg P ha^{-1}. Much greater values are reported for K, which is removed at rate increasing from 10 to 20 kg K per ha in this period. These values correspond to global deficits of 18.3 Mt N, 6.75 Mt P (50 Mt phosphate rock at a grade of 30% P_2O_5), and 30.56 Mt K (equivalent to 36 Mt K_2O). To compensate for these deficits would require an increase in world nitrogen fertiliser production of about 15%, and an increase in mining of

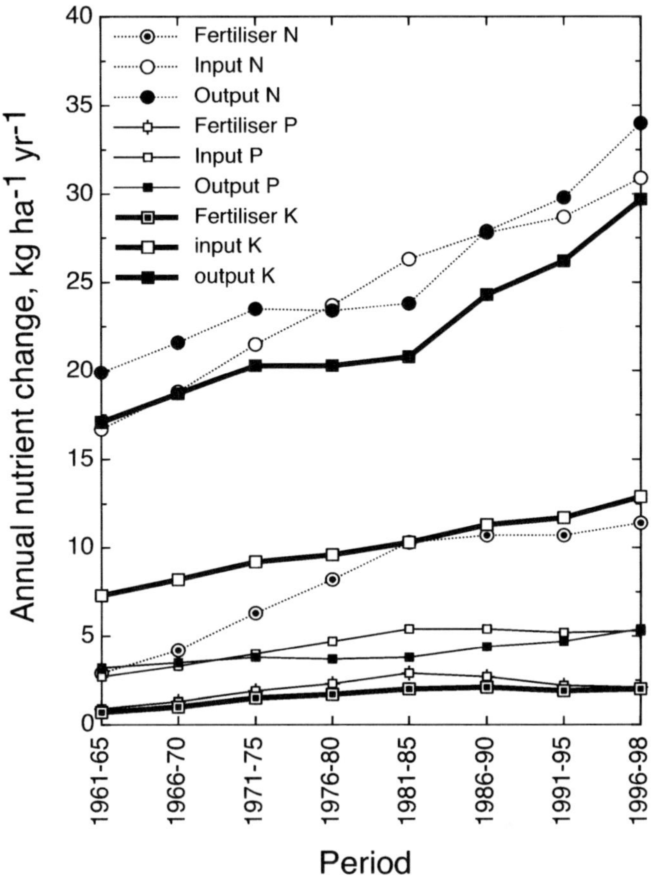

Figure 4 Nutrient balances for N, P and K for Africa.[17]

phosphate rock by 30–50%, depending on grade. In contrast, mined potash production would need to more than double to balance current agricultural requirements as expressed in terms of nutrient outputs.

4 Projections of Need to Support a Growing World Population

The nutrient audits carried out for 1996 correspond to a world population of 5.8 billion. In 2011, the world's population reached 7 billion, and is project to rise to 9 billion in 2050. In Africa, the current population is approximately 1 billion, and this will double by 2050. Food security is a major concern.[15,18]

Nutrient audits show that world production of fertilisers is insufficient to meet the needs of current populations. Simply on the basis of projected growth in numbers, the present deficit will only get worse; projected deficits are summarised in Table 4. This means that the world's arable soils will continue

Table 4 Estimated nutrient deficits for N, P and K, and corresponding amounts required to be mined (Ore), projected to 2050.

	2011		*2050*	
Population	*7 billion*		*9 billion*	
million tonnes:	*Element*	*Ore*	*Element*	*Ore*
N	19.51	-	28.40	-
P (grade 20% P_2O_5)	7.82	89.51	10.47	119.95
K	31.72	38.22	47.42	57.15

to be mined of nutrients, unless there are major changes in fertiliser production and supply. The scale of the problem is particularly severe for K, as production needs to triple by 2050 to meet projected global demand. In contrast, assuming a grade of 20% P_2O_5, world production of P needs to increase by about 70%, less if higher grades are mined.

5 Mineral Dissolution Rates in the Soil System

The demand for P and K fertilisers is currently met by apatite and soluble K salts, respectively. Although minerals such as feldspar can contain small amounts of P[19] (up to 1% P_2O_5), there is practically no widely available alternative to apatite. This means that the P deficit will be satisfied by the search for new resources and reserves of phosphate rock, to continue current mining practice. In contrast, there are many alternative sources of K, if we consider potassium silicate minerals. The difficulty with these is that they release their K slowly, by virtue of their slow dissolution rates.

Examples of K-bearing silicate minerals are given in Table 5, together with their typical K_2O content. Table 5 also shows dissolution rate data, for dissolution reactions involving an acid reaction mechanism.[20] Inspection of reaction rates shows that these vary by 9 orders of magnitude, which means that different potassium silicate minerals will behave very differently if used as sources of K.

Potassium feldspar occurs very widely in a range of igneous and sedimentary rocks, as one of three polymorphs: sanidine (typically from volcanic rocks), orthoclase (granites) and microcline (metamorphic rocks and syenites). It is one of the commonest minerals on the continental crust.[21] Having a log dissolution rate of -10.06 mol m^{-2} s^{-1}, it dissolves only very slowly.

Leucite and nepheline belong to the feldspathoid family; they occur typically in very specific volcanic rocks and in syenites. These rock types are uncommon, tending to occur in association with igneous rocks in rift valleys and other locations within continental plates.

Comparing the dissolution rates of feldspar, leucite and nepheline shows that there is a considerable difference, and this depends in detail on the

Table 5 Examples of K-bearing silicate minerals and their dissolution rates (acidic mechanism).[20] (Note that the K_2O content of nepheline varies from 3.6–12.2%. Dissolution rates for apatite and halite, as a proxy for sylvite (KCl), are also given).

Mineral	*Mineral family*	*Formula*	*Weight % K*	*Weight % K_2O*	*Dissolution rate, log mol m^{-2} s^{-1}*
Potassium feldspar	Feldspar	$KAlSi_3O_8$	14.0	16.9	−10.06
Leucite	Feldspathoid	$KAlSi_2O_6$	17.9	21.6	−6.00
Nepheline	Feldspathoid	$(Na,K)AlSiO_4$	8.3*	10.0	−2.73
Muscovite	Mica	$KAl_3Si_3O_{10}(OH)_2$	9.0	10.9	−11.85
Biotite	Mica	$K_2Fe_6Si_6Al_2O_{20}(OH)_4$	7.6	9.2	−9.84
Phlogopite	Mica	$K_2Mg_6Si_6Al_2O_{20}(OH)_4$	9.4	11.3	−10.00*
Hydroxyapatite	Apatite	$Ca_5(PO4)_3(OH)$	n.a.	n.a.	−4.29
Halite	Salt	NaCl	n.a.	n.a.	−0.21

*estimated

contrasting crystal structures of the three minerals. Remembering that the dissolution rates given in Table 5 are expressed as logarithms, the relative dissolution rates for the three minerals are such that leucite dissolves 10 000 times faster than feldspar, and nepheline 20 million times faster than feldspar.

Muscovite, biotite and phlogopite are micas, in which two tetrahedral Si-O sheets sandwich an octahedral sheet that contains Fe, Al or Mg coordinated with O. This package of three sheets carries a net negative charge, arising from the different valencies of Al, Fe, Mg and Si, and this is compensated by interlayer cations, typically K^+. The release of K from a mica occurs in two ways. On the one hand, a mica can dissolve, with dissolution rates corresponding to those given in Table 5. K can also be removed through cation-exchange processes, in which the structure of the mineral remains unaffected, while the interlayer cation is reversibly exchanged with a cation from a coexisting solution.

Table 5 also gives dissolution rates for apatite and, in the absence of data specifically referring to potash salts, for halite (NaCl). Apatite dissolves about 10^6 times more quickly than feldspar, and 10^4 times more slowly than halite. Overall, the range of dissolution rates given in Table 5 is 10^{10}, reflecting the different dissolution behaviour mechanisms of the covalently-bonded aluminosilicate structures typical of the feldspars through to the wholly ionic-bonded character of chloride salts of Na and K.

6 Possible Alternatives to Conventional Fertiliser Products

World food production depends on fertiliser inputs, and it is clear that insufficient is being mined to meet the needs of the present day. With world population projected to grow to 9 billion by 2050, fertiliser production needs to increase substantially, and that will require increased mining of the required raw materials. The need is especially great for potash; sustainable food production in 2050 is likely to require triple current mining activity; if increasing areas of land are used for biomass production for energy, this figure will increase accordingly.

Existing sources of P and K have substantial resources from which additional reserves can be defined, leading to continued mining of these for the foreseeable future. There is no doubt that phosphate rock and potash salts (Table 1) will continue to be of increasing importance as sources of soil fertility. But the increasing need for K, coupled with its very high price (which partly reflects the capital-intensive nature of mining salts underground), suggest that there is abundant scope for considering alternative sources of K.

A number of studies have considered the use of feldspars, feldspathoids and micas as sources of K.[22] In general, these have included crop trials intended to demonstrate whether or not a specific crop shows an increase in yield in response to application of a specific source of K. Statistical analysis of the results typically has shown that any observed response may be only marginally significant statistically,[23,24] and the value of a silicate mineral as a source of K has been dismissed. However, almost all of these studies were carried out when conventional K was a fifth of the price it reached in 2008.

The recent high price of K has led, in some countries, to a reduction in use of conventional fertilisers, simply because of cost, and renewed interest in alternative sources.[25] While the possible use of feldspars is of interest because of their high K content and widespread occurrence, the use of micas has also been investigated.

Experiments using zinnwaldite (8% K), a mica associated with tin mineralisation, and the feldspar orthoclase (10% K) have been carried out in the Czech Republic to investigate the response of barley.[25] As with many previous studies, the response to feldspar was not significant, but a statistically significant increase in yield was observed for treatments with zinnwaldite. In other studies to investigate olive tree growth,[26] a significant response was observed following application of glauconite (9% K_2O). In both of these studies, it is likely that the cation exchange reaction enabled these micas to act as sources of K to the soil solution.

Overall, the pathways by which minerals yield their nutrients to the soil solution for uptake by plants are summarised in Figure 5. The potassium silicate minerals, feldspars and feldspathoids, only release their K following weathering, and this typically produces K-bearing clays or micas. Micas yield K as a consequence of cation exchange reactions; they also undergo weathering, which tends to 'open up' the layered structure, facilitating release of K.[27] The potassium salts are directly soluble in the soil, with very rapid dissolution rates, resulting in K becoming immediately available to plants. The phosphate mineral apatite dissolves in soil systems, but 10 000 times more slowly than a simple salt, such as KCl.

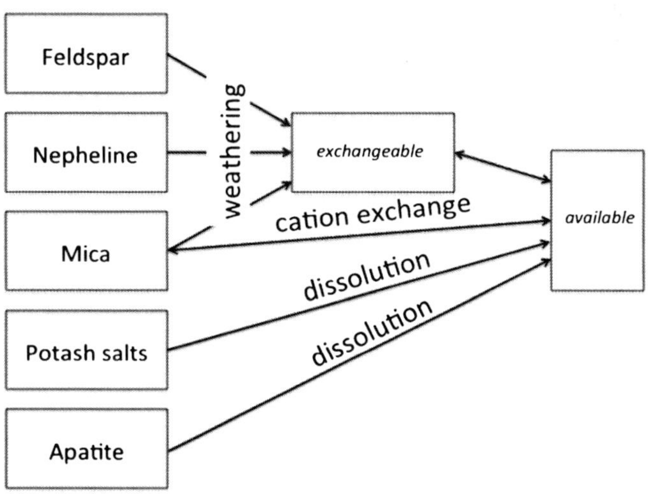

Figure 5 Summary of processes controlling release of nutrients from potassium silicate minerals, potash salts and the calcium phosphate mineral apatite to the soil solution, and so to a form available for plant uptake.

7 Conclusions

The world's population depends on adequate supplies of food. At the present time, food production is mining mineral nutrients from soil in an unsustainable way, especially for K. Much is known about the occurrence and quantities of both P and K that exist in the Earth's crust and can be mined as raw materials for fertiliser production. In the case of P, resources equivalent to 1800 years of production at current levels are known to exist. P availability is not a major issue, especially as so much P is lost through waste-water discharges that might otherwise be recovered. K has resources equivalent to 7500 years, but current production is half what is needed. This, in part, reflects the geographical and geological occurrence of K deposits, which occur mainly in North America and require production from capital-intensive deep mines. K is an expensive product, inaccessible on the basis of price to many farmers.

It is important to consider sources of K that represent an alternative to conventional expensive mined products. Potassium silicate minerals, such as feldspars and related minerals, are abundant and widely distributed globally. The key to their exploitation and use as sources of K is their dissolution rate, not their absolute K content. Additionally, micas also occur widely and provide a source of K that functions through cation exchange. It is reasonable to assume that necessity will lead farmers in poorer parts of the world to use silicate minerals as a source of K, in the absence of affordable alternatives.

References

1. United States Geological Survey, *2010 Minerals Yearbook: Nitrogen*, United States Geological Survey, Reston, Virginia, USA, 2011; http://minerals.usgs.gov/minerals/pubs/commodity/nitrogen/, last accessed 04/12.
2. United States Geological Survey, *2010 Minerals Yearbook: Phosphate Rock*, United States Geological Survey, Reston, Virginia, USA, 2011; http://minerals.usgs.gov/minerals/pubs/commodity/phosphate_rock/, last accessed 04/12.
3. United States Geological Survey, *2010 Minerals Yearbook: Potash*, United States Geological Survey, Reston, Virginia, USA, 2011; http://minerals.usgs.gov/minerals/pubs/commodity/potash/, last accessed 04/12.
4. British Geological Survey, *World Mineral Production 2006–2010*, British Geological Survey, Keyworth, Nottingham, 2012.
5. M. Lægrid, O. C. Bøckman, and O. Kaarstad, *Agriculture, Fertilizers and the Environment*. CABI Publishing, Wallingford, 1999.
6. Food Standards Agency, *McCance and Widdowson's The Composition of Foods*, Royal Society of Chemistry, 6th Summary edn, 2002.
7. J. D. Jones, P. B. Kaufman and W. L. Rigot, *J. Radioanal. Chem.*, 1979, **50**, 261.
8. Department for Environment, Food and Rural Affairs, *Fertiliser Manual (RB209)*, 8th edn, 2010.

9. United States Department of Agriculture, *Thailand: Sugar Semi Annual. Global Agricultural Information Network Report Number TH1123*, 2011; http://gain.fas.usda.gov, last accessed 04/12.
10. United States Geological Survey, *Mineral Commodities Summary, Appendices*, United States Geological Survey, Reston, Virginia, USA, 2011; http://minerals.usgs.gov/minerals/pubs/mcs/2012/mcsapp2012.pdf, last accessed 04/12.
11. Food and Agriculture Organization of the United Nations FAO, Current world fertilizer trends and outlook to 2013/14. 2010; ftp://ftp.fao.org/ag/agp/docs/cwfto14.pdf, last accessed 04/12.
12. D. Cordell, J. -O. Drangerta and S. White, *Global Environ. Change*, 2009, **19**, 292.
13. R. G. Ponce and M. E. G. L. De Sa, *Agrochimica*, 2007, **51**, 301.
14. G. K. MacDonald, E. M. Bennett, P. A. Potter, and N. Ramankutty, *Proc. Natl. Acad. Sci. U. S. A.*, 2011, **108**, 3086.
15. P. M. Vitousek, R. Naylor, T. Crews, B. David, L. E. Drinkwater, E. Holland, P. J. Johnes, J. Katzenberger, A. Martinelli, A. Matson, G. Nziguheba, D. Oima, C. A. Palm, G. P. Robertson, P. A. Sanchez, A. R. Townsend and F. S. Zhang, *Science*, 2009, **324**, 1519.
16. W. F. Sheldrick, J. K. Syers and J. Lingard, *Nutr. Cycling Agroecosyst.*, 2002, **62**, 61.
17. W. F. Sheldrick and J. Lingard, *Food Policy*, 2004, **29**, 61.
18. P. J. Gregory and T. S. George, *J. Exp. Bot.*, 2011, **62**, 5233.
19. D. London, P. Černý, J. L. Loomis and J. J. Pan, *Can. Mineral.*, 1990, **28**, 771.
20. J. L. Palandri and Y. K. Kharaka, *U.S. Geological Survey Open File Report 2004–1068*, Menlo Park, CA..
21. D. A. C. Manning, *Elements*, 2008, **4**, 105.
22. D. A. C. Manning, *Agron. Sustainable Dev.*, 2010, **30**, 281.
23. A. K. Bakken, H. Gautneb, T. Sveistrup and K. Myhr, *Nutr. Cycling Agroecosyst.*, 2000, **56**, 53.
24. A. D. Harley and R. J. Gilkes, *Nutr. Cycling Agroecosyst.*, 2000, **56**, 11.
25. M. Madaras, M. Mayerová, M. Kulhánek, M. Koubová and M. Faltus, *Arch. Agron. Soil Sci.*, in press.
26. E. Karimi, A. Abdolzadeh, H. R. Sadeghipour, and A. Aminei, *Arch. Agron. Soil Sci.*, 2011; DOI: 10.1080/03650340.2011.557369.
27. A. Psyrillos, J. H. Howe, D. A. C. Manning, and S. D. Burley, *Clay Miner.*, 1999, **34**, 193.

Soil Physical Degradation: Threats and Opportunities to Food Security

PAUL D. HALLETT*[a], KENNETH W. LOADES[a] AND JULIA KRÜMMELBEIN[b]

[a] The James Hutton Institute, Invergowrie, Dundee, DD2 5DA, United Kingdom; [b] Lehrstuhl Geopedologie und Landschaftsentwicklung, Brandenburgische Technische Universität, Konrad-Wachsmann-Allee 6, 03046 Cottbus, Germany
*E-mail: paul.hallett@hutton.ac.uk

ABSTRACT

Soil physical degradation affects over 1000 million ha of land globally. It includes compaction, erosion, and the slumping, capping and hardsetting of seedbeds. Trends in recent years are promising. The widespread adoption of reduced tillage systems has decreased erosion and other physical stresses on soils considerably in some regions. New technologies such as satellite navigation and Controlled Traffic Farming offer potential to decrease soil damage by limiting wheelings to specific locations.

Nevertheless, the threat of soil physical damage continues and many soils are already degraded. Soil compaction is a recognised threat that could worsen given increasing trends in machinery weight. Simple problems such as the deterioration of seedbed physical structure after

Issues in Environmental Science and Technology No. 35
Soils and Food Security
Edited by R.E. Hester and R.M. Harrison

Published by the Royal Society of Chemistry, www.rsc.org

cultivation remains poorly understood. Without detailed knowledge of soil physical conditions and how they interact with crops, optimising crop yields is not possible.
There are serious implications to food security. Soil erosion and capping can render large areas unproductive for agriculture. Modern crops have been bred for ideal soil conditions that do not exist on many farms. Selecting root traits that can penetrate compacted layers, access biopores and form an extensive network to access water and nutrients requires collaboration between plant and soil scientists. There is considerable untapped potential beneath our feet that needs to be tapped into to secure the global supply of food into the future.

1 Introduction

Crop yield and hence food security are limited considerably by physical constraints from soil.[1] Farmers recognise the huge yield gap they face due to physically constrained soils, so over the past several decades major changes have occurred in soil management practices.[2–4] Since Roman times, the plough dominated agricultural production systems. Combined with secondary tillage operations, ploughing was used to break up compacted layers and form an aggregated seedbed that was ideal for plant establishment and subsequent growth.[5] For many regions of the globe, particularly in higher yielding temperate regions, the use of the plough continues. It is viewed as an essential intervention to maximise yield and hence farm-gate profitability. Whether this is based on fact or habit will be discussed in this chapter.

Intensive farming and ploughing, however, have had a negative impact on many farming regions internationally.[6,7] At the extreme end are environmental disasters, including the Great Dust Bowl in the US Midwest in the 1930s, and current dust bowls that plague the Loess Plateau region in China. Ploughing, intensive grazing and changes in vegetation stripped the soils in these areas of organic matter and fibrous root systems that held the soil together. Large-scale erosion and dust storms resulted, producing crop failures locally and dust storms that have been tracked from China to the west coast of the USA.

Farming practices can also compact soil from machinery traffic or animal grazing.[8] Compaction decreases the amount and connectivity of soil porosity, so the transport of water and gases through soil and available spaces for root growth and microbial colonisation becomes impaired.[1] As compacted soils dry, the strength that develops can impede root growth and hence the capacity of plants to access water and nutrients from deeper soil layers.[9] Yield penalties of up to 30% are not uncommon in compacted soils.[10]

Another major soil physical constraint to crop production is soil structural instability. Crusting of surface soils from rainfall can produce barriers to seedling emergence.[11] Unstable seedbeds also slump over time, resulting in the coalescence of previously aggregated structures into a denser mass.[12] Splash impacts from rainfall on unstable soils, coupled with overland flow caused by diminished infiltration rates, erode soil and result in soil deposition in either lower-lying soil areas or in water bodies. Nutrients carried with these sediments, particularly phosphorus, can cause eutrophication of water bodies.

Practices other than agriculture can cause physical disturbance to soils. Construction engineering and mining activities access deeper soil layers to establish foundations, install infrastructure or access buried resources. Although these activities can cause severe soil physical degradation, this can be managed through careful control of machinery traffic and the stockpiling of soil to preserve topsoil and avoid mixing. Strict legislation exists in many countries requiring operators to 'recultivate' soils so that they can be used for agricultural production once mining projects are complete or cease. A scientific discipline has emerged around this topic that investigates best practice to avoid and mitigate soil physical damage.

The attention paid to soil physical degradation in the context of food security varies considerably between regions. Severe soil physical degradation is visually apparent and widespread on sensitive soils that unfortunately make up the natural capital of many developing nations.[13] Efforts to mitigate the problem in these regions are widespread where funds and farmer incentives are made available. In higher quality soils that are responsible for producing a large proportion of the global food supply, soil physical degradation is recognised but not given the attention it often deserves. An imbalance of research interest to address soil physical degradation has resulted in recent years due to a shift in investment towards agricultural biotechnology. Focus on the plant has forgotten about the soil.

There is considerable scope, however, to mitigate physically damaged soil to enhance crop yield and in many cases also decrease the environmental impact of farming. A large-scale shift in farming practice towards minimum tillage, where decreased physical disturbance increases surface carbon levels and biology, is one approach to improve soil physical conditions.[14,15] Successful farming in the Cerrado Region of Brazil, for instance, was only possible because of this change in soil management practice. These changes were spearheaded by the scientists and agricultural advisors Colin McClung, Edson Lobato and Alysson Paolinielli, who received the World Food Prize in 2006 for helping to unlock Brazil's massive potential for crop production.[16] Other approaches to mitigate or avoid physical damage to soil include controlling the traffic of vehicles to dedicated tramlines, decreasing the stocking density of livestock, amending soils with organic matter, sub-soiling to remove compacted pans that form beneath the plough and the use of amendments to stabilise soils or improve water flow.

In this chapter we review and discuss the challenges and opportunities of soil physical degradation. An overview of the forms of degradation and global implications is provided first, followed by the potential opportunities provided by practices and technologies to avoid or mitigate the problem. Associated challenges such as the need for decreased fuel, fertiliser and water inputs are taken into account. In some instances, such as the adoption of reduced tillage systems, win–wins of decreased inputs and decreased soil physical damage can be achieved.

2 Forms and Extent of Soil Physical Degradation

Quantifying the extent of soil physical degradation is not an easy task. The 'Global Land Assessment of Soil Degradation' (GLASOD) was setup in the early 1990s to attempt to provide expert advice on the current situation. Human-induced soil degradation has affected 24% of the global inhabited land area, 1966 Mha. The global range of degradation ranges by continent from as low as 12% in North America to 31% in Asia. Within the term of soil 'physical degradation' used by GLASOD, three types exist: (i) compaction, crusting and sealing; (ii) waterlogging; and (iii) subsidence of organic soils. Compaction, crusting and sealing have caused an estimated 4% (68.3 Mha) of degradation, with waterlogging and subsidence of organic soils degrading 11 Mha and 4 Mha, respectively. Of all soil physical degradation, agricultural mismanagement has accounted for 80% and overgrazing 16%.

The large areas of the globe affected by soil degradation are shown in Figure 1. By far the greatest form of degradation is erosion by wind and water, with pockets of easily physically degraded soils in isolated regions. The quality of data used to map soil resources, however, was discussed in Chapter 4. There is insufficient evidence available to accurately map many of these threats and the map itself is over 20 years old. Over the past two decades a major negative change has been increased machinery weight, but positive changes also have occurred, such as better organic management, reduced tillage adoption in sensitive regions, and new engineering technologies that decrease damage. An interesting observation from Figure 1 is that the physically degraded region in Sweden corresponds to a country where considerable research investment in soil compaction occurs. Although the threat of soil compaction may have encouraged greater research investment, it is likely that the soils of other regions, such as the UK, are also vulnerable, but poorly characterised.

Figure 1 only shows the major form of soil degradation. Different types of soil physical degradation often interact. Soil compaction, hardsetting and slumping all can enhance soil erosion. A management practice designed to decrease one form of soil physical degradation generally has a positive impact on other types of degradation. Organic matter incorporation, for instance, can decrease the slumping of seedbeds, erosion by either wind or water, and the resilience of soil to compaction.

2.1 Soil Compaction

Soil compaction is the loss of porosity through mechanical damage to soil. This definition can be extended to encompass all forms of mechanical damage that decrease crop productivity or have a negative environmental impact.

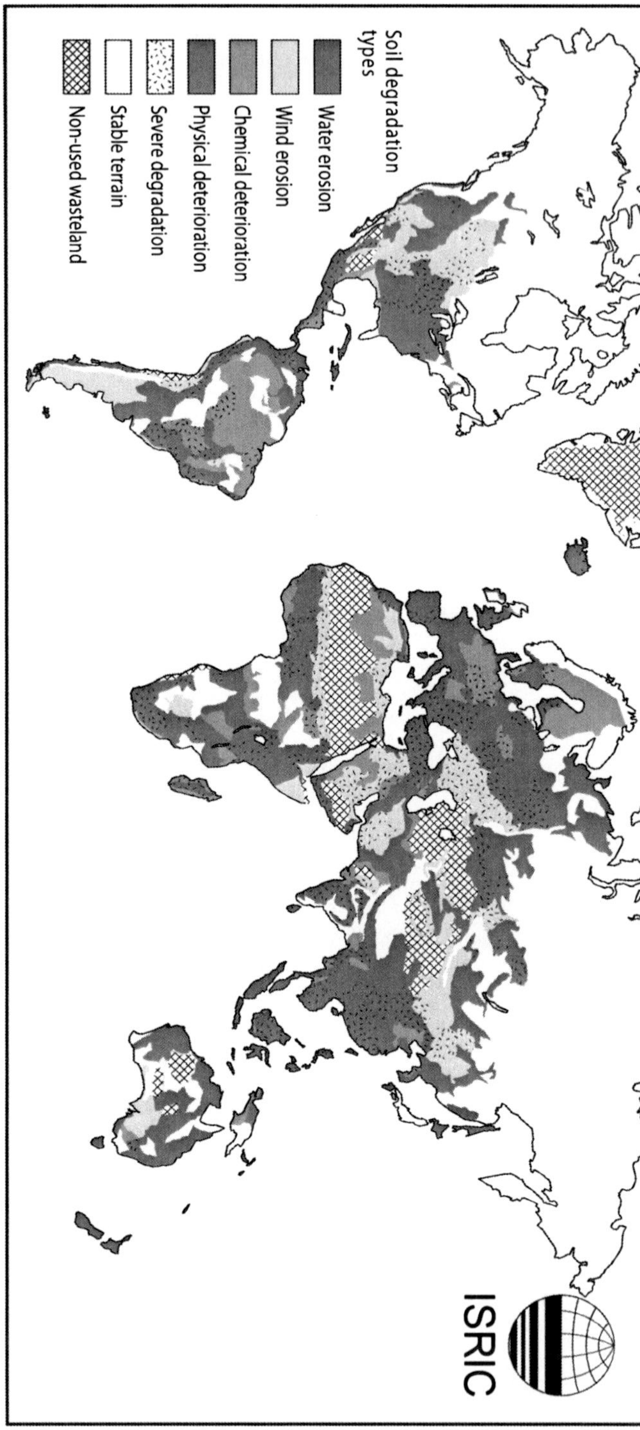

Figure 1 Human-induced soil degradation estimated across the globe. (Source: Oldeman *et al.*).[35]

Topsoil compaction refers to damage to the ploughed layer of soil where remediation or natural processes can assist with structural recovery following damage. *Subsoil compaction* refers to damage below the plough layer, where remedial measures to alleviate structural damage are not possible or are less effective, resulting in more persistent damage. Soil compaction is classified as harmful when saturated hydraulic conductivity and air capacity go below certain values (<10 m day^{-1} and 5%, respectively).[17]

Soil compaction occurs when an external mechanical stress from equipment or livestock exceeds the mechanical stability of soil.[18] Many factors control the susceptibility of soil to compaction, including the stress history from previous traffic,[19] texture,[20] organic matter[21] and soil structure.[22] Most of these properties also affect how well soil recovers from compaction through subsequent cultivation[23] or the inherent resilience under natural weathering.[24]

The amount of water in soil affects its mechanical stability to compaction. As a consequence, precipitation changes predicted for the future could have major implications.[25] Drier soils compact less than wetter soils. In the European Union, the greening of the Common Agricultural Policy (CAP) through the implementation of the Good Agricultural and Environmental Code (GAEC) restricts access to fields when they are wet.[26] However, providing accurate advice to farmers on when not to access land with machinery or livestock is a challenge. Predicting the susceptibility of soil to compaction is fraught with uncertainty because of the complexity of stress transmission through soil and confounding factors such as the influence of soil structure.[27]

Subsoil compaction is viewed as a far greater threat[28] than topsoil compaction, as the latter can be ameliorated to some extent by soil cultivation and natural processes.[21,29] The impacts of subsoil compaction include reduced crop yields, poor drainage and increased overland flow.[30–32] Compaction also alters microbial habitats in soil,[24] so it could have implications for biodiversity.[33]

2.1.1 Extent of Compaction

Evidence of soil compaction is extensive globally, particularly under intensive agriculture where machinery weights are large.[28,34] About 4% (69.3 Mha) of anthropogenic soil degradation globally has been attributed to soil compaction,[35] but this estimate is over 20 years old and limited to extreme compaction where effects are evident from crop failure or severe surface damage. An analysis of soil on 156 farms in Scotland by Ball[36] found that many had been affected by soil compaction. This study followed on from anecdotal evidence for Scotland collected by Soane,[37] who found widespread concern about soil compaction from 1421 farmers who were surveyed.

Soil compaction can be out of sight and out of mind. The effects are not always evident and sometimes remedied unknowingly through the increased application of fertilisers. Simply by managing soil to have a smaller bulk density, Soane & Van Ouwerkerk[38] demonstrated that applied N can be cut by almost 30% and maintain output. Other research has shown up to 6% increases in wheat yield in the absence of soil compaction.[39] Soil compaction increases the penetration resistance of soil, which is a major limitation to crop productivity.[9,40]

The impact of soil compaction on crop yield varies considerably between soil types and regions. A recent study applied 'light' (1 pass) and 'heavy' (8 passes) compaction stresses to soils with different textures and then sowed the fields with wheat.[24] The yield reduction due to 'heavy' soil compaction in a sandy loam soil was almost 50%, whereas for a clay soil the yield was not changed by applying 'heavy' compaction stresses. The authors attributed the results found for the clay soil on the buoyancy effect of pore water. In a review of many studies examining soil compaction impacts on yields, Chamen[10] found a large variability in yield decrease caused by compaction. In comparisons of trafficked *vs.* non-trafficked soils, 15% yield decreases were quite common, but unstable soils could have yield decreases greater that 40%, whilst some crops and soils had increased yield following compaction.

Figure 2 shows a conceptual diagram of the implications of soil compaction to agricultural productivity and the environment. Decreased water infiltration caused by a reduction in pore space leads to greater surface run-off of water

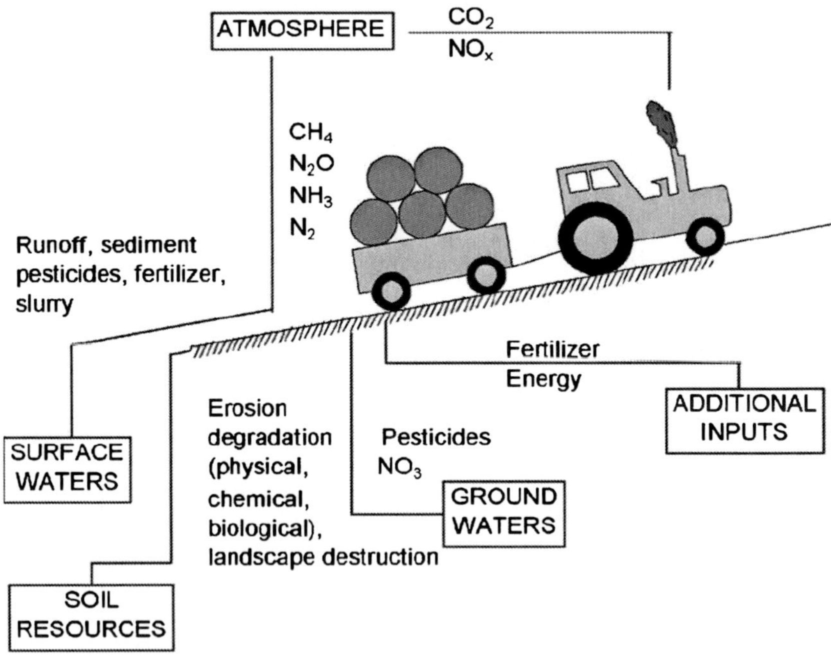

Figure 2 A conceptual diagram that shows the various implications of soil compaction to the environment. The potentially negative implications to crop productivity are not shown. (Source: adapted from Soane and Vanouwerkerk).[37]

and therefore to erosion. The presence of tramlines, where multiple wheelings of tractors occur from spraying and other management practices, can cause a 10-fold increase in the amount of transported phosphorus that is carried on soil sediments.[41] In the UK alone, agriculture accounts for 20% of phosphorus pollution to surface waters, so mitigation practices to decrease compaction would have great environmental benefits. As soil compaction increases, soil aeration and waterlogging become more problematic, leading to increased greenhouse gas production. Soil biological habitats are also disrupted by soil compaction, with impacts evident for earthworms[42] but less so for microbial diversity.[43]

Given trends in increased machinery weight over the past several decades,[44] soil compaction could worsen. Root-crop harvesting equipment now weighs in excess of 40 tonnes, which is beyond the legal limit for many roads. Increased machinery weight causes greater stress transmission to depth in soil, so damage to the subsoil becomes more problematic. Persistent damage of the subsoil has been estimated to occur for wheel loads greater than 3–4 tonnes.[45] The weight of machinery is linked to machinery power, which has increased seven-fold in Germany since 1950.[44]

Limited research has considered the potential impacts of climate change. Where precipitation is predicted to decrease, such as in Mediterranean regions, drier soils should decrease the occurrence of soil compaction. However, already-compacted soils store less water, so the implications for crop production will be exacerbated. Good timing of soil management operations is considered to be essential to protect soil vulnerable to compaction.[46] In Scotland, Cooper *et al.*[25] predicted that climate change will lead to a marked reduction in days when soil is not at risk of compaction because of wetness. Climate change is anticipated to lead to more erratic weather patterns, thus diminishing the number of consecutive days available to farmers for field operations.

2.2 Soil Erosion

Soil erosion can range from large-scale mass-wastage events, such as landslides, to smaller scale events driven by the gradual spatial displacement of soil particles by the action of wind or water. Both wind and water erosion occur when the bonds between soil particles or small soil aggregates are too weak to resist detachment and entrainment from the forces of air or water. Mechanisms involved in water erosion include: (i) raindrop or splash impact; (ii) overland flow of water; and (iii) slaking of aggregated soil structures through rapid wetting.[47,48] Tillage erosion refers to the movement of soil down slopes through gravity during soil cultivation.[49]

The erosion of soils is a natural process that underlies the formation of landscapes. Providing that the rate of soil loss does not exceed the rate of soil formation, it is not a problem. Since the beginning of agriculture, however, human impacts from soil disruption through tillage or trampling by livestock have exacerbated rates of erosion.[5] Various properties of soil influence its resistance to erosion. Surface cover by vegetation decreases the impact of raindrops on soil and buffers overland flow.[50] Soil particles are enmeshed by roots and fungal hyphae that are broken by mechanical disruption from soil cultivation.[51] Compounds exuded by soil biota increase the adhesion and hence aggregation of soil particles.[52,53] Rearranging soil particles and stirring in oxygen by cultivation increases the rate at which organic compounds are mineralised, resulting in decreased amounts of soil organic carbon.[54] Erosion resistance and soil organic carbon are closely correlated properties of soil.

2.2.1 Extent of Erosion

Agriculture accounts for about three quarters of soil erosion globally. As Figure 1 illustrates, the problem is widespread, with about 80% of agricultural land plagued by moderate-to-severe erosion.[50] These figures are worrying and have resulted in the phrase 'Peak Soil' coming into use. Civilisations have fallen due to soil erosion. From 4000–2000 BC, Uruk in southern Mesopotamia had up to 80 000 residents and was the first recorded city.[55] Overgrazing by livestock and forestry prompted erosion and poor soil drainage. Flooding was commonplace, with extreme events turning vast areas of the surrounding terrain into a sea with the city serving as a refuge. It was the birth of the story of Utnapishtim, a flood hero similar to Noah. By 2000 BC, food shortages due to soil erosion and salinisation caused Uruk to become uninhabitable. The inhabitants of Easter Island, the Mayans and others had a similar fate.

In modern agriculture, a global compilation of erosion studies found that ploughed fields eroded 1–2 orders of magnitude more rapidly than soil was formed.[13] The human-induced impact caused by soil cultivation and vegetation removal was evident from this broad analysis of available data. Cultivating soil for agriculture caused a median increase in soil erosion that was 18 times greater than the level experienced under natural vegetation. The rate of soil loss *versus* soil formation has been characterised to predict longer term trends in soil sustainability. In Europe the upper limit of soil erosion based on formation rates ranges from 0.3 to 1.4 t ha^{-1} yr^{-1}.[56] The actual level of soil erosion in Europe, however, is 3 to 40 t ha^{-1} yr^{-1}, which is comparable to global estimates of soil loss rates at 10–40 times greater than the rate of soil formation.[50]

Of the 1966 Mha of land estimated to be physically degraded, water erosion is responsible for 1100 Mha and wind erosion for 550 Mha. Extrapolating the degree of degradation further, approximately 225 Mha of soil has been degraded by water erosion to such an extent that it is unsuitable for agriculture. It is estimated that the primary cause of soil lost through wind erosion results from overgrazing, accounting for 60% of soil losses.

The stripping of soils of vegetative cover for construction or mining activities can cause erosion of soil from 20 to 500 t ha^{-1} yr^{-1}.[50] Most countries have strict legislation in place to quickly restore affected land, which is also in the interest of contractors so that mitigation costs are minimised.

Soil erosion rates are decreasing due to changes in soil cultivation practice, abandonment of agriculture on extremely vulnerable soils and through the maintenance of crop stubble or cover over the winter. Climate change is predicted to have a variable impact on soil erosion. With available climate models, such as the UK Meteorological Office's Hadley Centre HadCM3 Global Circulation Model, the spatial and temporal resolutions of precipitation data are inadequate for erosion models.[57] Downscaling approaches allow estimates to be made that suggest increased erosion in some regions, primarily due to high rainfall intensity events, but decreased erosion in other areas.[57–59] Decreased total precipitation and increased biomass are the two main factors that could decrease erosion in the future because of climate change.[57] In a climate hindcasting study, erosion rates in the severely affected Mediterranean badlands were shown to have decreased from 1974 to 2004 because of decreasing precipitation.[59]

Although soil erosion is a recognised threat that has prompted large changes in soil management, the need to produce more food in some of the worst affected regions could exacerbate erosion rates.[60] In developed nations, government policies such as the greening of the Common Agricultural Policy to pay incentives to farmers for good stewardship help address the problem. Smallholders in developing countries, however, often do not have access to information or technologies. Large-scale programmes such as AGRA[61] aim to deliver soil conservation, food security and poverty reduction simultaneously.

2.3 Seedbed Instability

The primary purpose of soil cultivation is to optimise a seedbed to maximise crop productivity. A tilled seedbed contains more than 50% of the root mass of a developed plant, so it is extremely important to crop yield.[62] Over time, however, the aggregated structure produced by cultivation can be disrupted by slaking or drop impact from rainfall,[48] or coalesce and slump[63] under its own self-weight and mechanical instability, resulting in poorer soil physical conditions (Figure 3). Seedbed instability manifests itself in various forms, including (i) surface crusting, (ii) hard-setting and (iii) the slumping of structure. The consequences for crop production include increased risks of hypoxia, poorer water storage and greater mechanical impedance to seedling emergence and root growth.[11] Unstable seedbeds are also more prone to erosion and damage from compaction.

It is widely appreciated that changes to seedbed structure over time remains one of the least well understood processes in soil.[12] Considerable research exists that monitors changes in soil structural properties over time, but very little research has examined the underlying mechanisms or attempted to develop predictive models.[64] As with soil erosion, decreased amounts of carbon in soil can result in seedbeds that deteriorate rapidly post

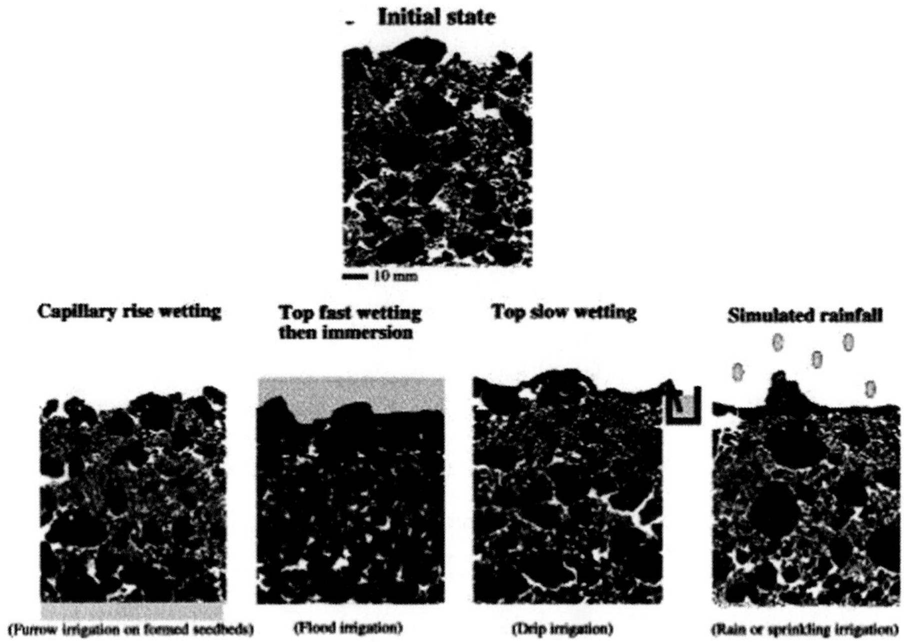

Figure 3 Binary images of vertical slices of soil (pixel size 107 μm, solid in black): initial seedbed, bottom slow wetting (BS), top fast wetting then immersion (TFI), top slow wetting (TS) and rainfall simulation (R). (Source: Bresson).[2]

cultivation.[65,66] It has been known for decades that the slaking and coalescence of soil structure increases if bonding by organic compounds is diminished.[67]

2.3.1 Extent of Seedbed Damage

Hardsetting of soils poses a major risk to crop production in the sub-humid to semi-arid tropics and Mediterranean regions where it is prominent. It affects more than 110 Mha of agricultural land. Affected soils, once cultivated, disperse under rainfall and then dry into dense, structureless layers that are difficult for water or plant roots to penetrate.[68] Many studies suggest that a shift to reduced tillage systems can decrease hardsetting over time, but the longer-term impacts are not known.[69]

Surface crusting and sealing affects all regions of the globe, but is most prominent in Africa (18 Mha) and Asia (10 Mha).[35] Declines in organic matter caused by intensive cultivation and trampling by livestock are the primary causes.

Seedbed slumping occurs to some extent in all soils following tillage.[70] It is very poorly characterised and only appreciated when associated with water-logging, which affects about 11 Mha worldwide.[35] The maintenance of the physical structure of seedbeds, however, is vital to crop productivity and offers considerable untapped potential in the quest to increase crop yields. Seedbed slumping not only decreases the storage of water and transport of gases that are essential for crop growth, it also increases the strength of soil upon drying.[9] The impedance to root growth caused by soil strength increases from drying has a major impact on yield.[71] Soil compaction exacerbates the problem further.

3 Measuring Soil Physical Constraints

Crop productivity is limited by the following soil physical constraints:

- Hypoxia – depletion of oxygen at the root–soil interface. The oxygen concentration or redox potential of soil can be measured directly to assess hypoxia. Often a value of 10% air-filled porosity is assumed to be the cut-off for severe hypoxia to occur.[72] This is assessed from the water content and porosity of soil.
- Water potential – capillary stresses of pore water that create suction in soil. As suction increases, plants have to overcome a greater stress to extract water from soil. A water potential of −1500 kPa defines the permanent wilting point of soils, where crops can no longer access water, leading to wilting and death.[1] Water potential is measured with tensiometers in the field or from water-release characteristics obtained using pressure-plate apparatus in the laboratory.
- Mechanical impedance – roots push their way through the soil matrix to increase the volume of soil they access and hence the capture of water and nutrients. Increased soil suctions (drying), compaction and seedbed slumping can increase the mechanical impedance of soil. At a threshold value of 2 MPa, root elongation is impeded severely.[71] Mechanical impedance is measured with a penetrometer, either in the field or the laboratory under more controlled conditions.
- Macropores – these larger pores in soil provide rapid transmission pathways for gases and water, in addition to continuous void space for root growth.[73] They are formed either by the cracking and aggregation of soil through weathering or by the action of soil biology, such as earthworms or plant roots.[33,72] Macropores are measured either from water-release characteristics (macropores drain at small suctions) or through visual methods such as thin sections or X-ray computer tomography.[1]

Although cut-off values for root growth and functioning have been provided above, the values are not fixed and the different physical constraints are interdependent.[74] This is described in greater detail below.

Internationally there has been a recent surge in the development of soil-quality indicators to measure physical constraints to crop productivity. Many are based on the distribution of different pore classes in soil, as these provide indicators of oxygen exchange and water transport to prevent hypoxia, water storage to resist drought, and the presence of macropores to provide preferential channels for root growth.[1,73,75,76] A simple indicator, the *S index*, has been developed that uses the shape of the water retention curve, which is the relationship between soil suction (from capillary forces) and water content.[77] The *S index* provides a numeric value that has been correlated to fragmentation of soil by tillage, seedbed degradation[78] and the optimal water content for timing of tillage operations.[79] This approach is being adopted in other regions,[80] with over 150 citations[77] from other field studies, but its theoretical basis has raised questions and verification against crop performance is needed.

A more commonly used indicator to identify soil constraints to crop productivity is the least-limiting water range, LLWR.[81,82] It uses cut-off values of soil water content based on hypoxia, drought and mechanical impedance to define the range of water contents where soil properties will not severely impede crop productivity (Figure 4). LLWR is based on several generalisations about soil behaviour that may not be observed in field conditions, where the structural heterogeneity of soil properties can have a large impact on crop productivity. For instance, LLWR defines 2 MPa as the cut-off mechanical impedance where root growth is completely restricted. Field penetrometer resistance that limits oat root growth, however, was found to be

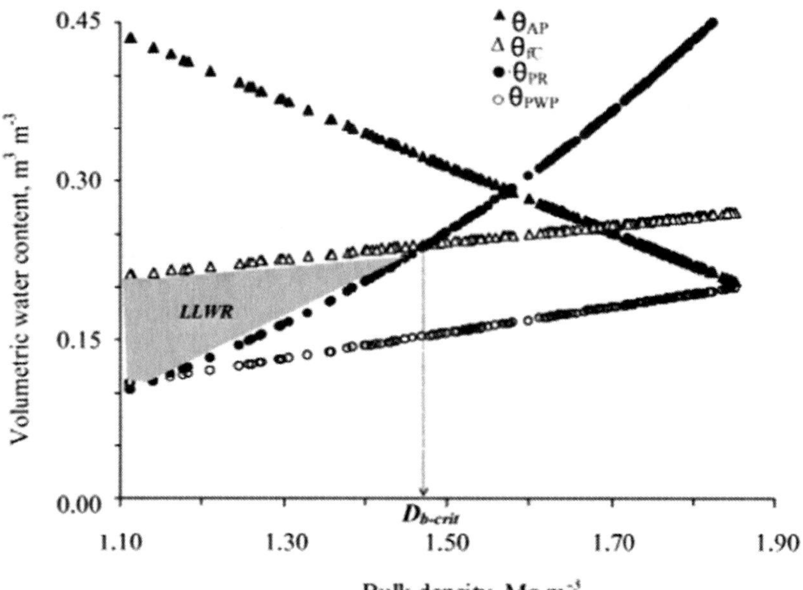

Figure 4 Concept of the Least Limiting Water Range (LLWR), showing an example of a soil from Brazil-growing soybeans. Plants are estimated to become severely limited if the air-filled porosity, θ_{AP} <0.10 m^3 m^{-3}, the soil is wetter than field capacity, θ_{FC}, the penetration resistance, θ_{PR}, is greater than 2 MPa and at the water content of permanent wilting point, θ_{PWP}. (Source: Beutler *et al.*).[81]

4.6 to 5.1 MPa in untilled soil layers, as compared with 3.6 MPa for tilled topsoils[83] due to roots exploiting a network of continuous biopores in the tilled soil. Such pore networks enable roots to penetrate to depth in very hard Australian sub-soils, where wheat-root growth is confined almost entirely to biopores.[84] Plenty of scope therefore exists to develop better soil quality indicators that describe limitations to crop productivity.

Root penetration rates have been used in the field[85] and laboratory[86] to identify soil and plant limitations to rooting depth. These show promise in matching genotype to farm management, thereby offering considerable potential to increase productivity.[87] Crop cultivars perform differently under a range of tillage practices[88] due to the rate and formation of root establishment and tolerance to soil physical stress. Phosphorus-use efficiency in soils with different physical constraints can also vary significantly between different genotypes.[89] Cultivar performance in relation to soil biopores is also being investigated using a new screening approach.[90]

4 Soil Physical Restoration and Food Security

Crop yields can be enhanced by improving soil physical conditions and selecting crop traits that are more resilient to soil physical constraints.[1] Despite impressive improvements in the yield of key crops globally over the past several decades, achieving even greater yield presents a great challenge.[91] The focus in plant science research has been above ground, in the leaves, stems and crop. A switch is occurring to consider plant roots in greater detail, because it is recognised that this is where the untapped potential lies. Chapter 1 describes this research in greater detail. By selecting root traits that are resilient to the physical constraints imposed by soil, greater yield should be possible, particularly in drought-stricken regions.[84]

Livestock production is another facet of food security that will benefit from improved soil physical conditions.[92] Earlier, the effects of overgrazing on soil physical conditions, particularly erosion, were reviewed. Compacted or 'poached' fields produce less vegetation and hence less feed for livestock. There is potential both for improved management and for improved pasture crop varieties so that production systems become more sustainable.

There are some impressive global examples where improvements to the management of soil physical conditions have underpinned local food security and successful farming enterprises. Arable farming would be much less viable in the Cerrado region of Brazil [93] and many regions of Australia[94] if reduced tillage farming had not been adopted locally. Food is now produced on what used to be severely degraded soils in the Loess Plateau and Red Soil region of China.

A range of practices exist to manage or improve the physical condition of soil. The massive shift in soil cultivation systems towards reduced tillage has been the greatest change globally. Compacted soils can also be subsoiled and tilled intensively to break up plough pans, but the longer-term effectiveness is questionable. On unstable or severely degraded soils, changes in vegetation can restore soil fertility and provide enmeshment and reinforcement to prevent erosion and landslides of slopes. New technologies have been developed to avoid soil physical damage. These range from tyres or tracks that decrease topsoil compaction, to sophisticated satellite navigation systems to minimise trafficked areas on fields.

4.1 Soil Cultivation

4.1.1 Subsoiling

Subsoiling is the mechanical disruption of plough pans beneath the normal depth of cultivation. It remains in common use in many regions of the world and is viewed as an essential practice by some farmers. The evidence on its effectiveness is mixed. Marks and Soane[95] measured crop yield at 25 sites in the UK and found that on 75% of the sites subsoiling had a neutral to negative impact. Soils that had been subsoiled may be more susceptible to subsequent compaction damage by machinery. Chamen[10] found a neutral effect of subsoiling on yield as well, and found that two tractor passes could return soil to a state where it was stronger than its original condition. Given the poor effectiveness of subsoiling, producing isolated fissures in extremely compacted regions (*e.g.* tramlines) has been advocated instead of large-scale 'loosening' of entire fields.[96]

When attempting to restore a compacted soil by subsoiling it is important that the soil is dry enough and does not deform plastically. Otherwise the subsoiling device does not break the soil and create new pores (cracks) but smears through it, creating thin planes with even lower conductivities for gases and liquids. For deep subsoiling (>60 cm) it is doubtful that in most temperate regions that these depths are ever dry enough for subsoiling for more than a few days per year.

Another factor determining the persistence of subsoiling is the internal strength of the soil against mechanical stresses, which is significantly decreased by subsoiling.[97] Soils with low structure formation intensity, *e.g.* sandy soils with small amounts of organic matter or other stabilising substances, show the least persistence of subsoiling. It is widely thought that a minimum content of clay of about 25% is needed for persistent deep loosening, although Marks and Soane[95] challenge this view. One measure to support the stabilisation of newly created pores from subsoiling is the cultivation of plants with fast- and deep-growing root systems that quickly reach and penetrate fractures. Subsoiling, although widely in use, appears ineffective and even damaging to soil physical conditions for crop growth in the long term.

4.1.2 Shift to Lower Input Systems

Agriculture has undergone a revolution in soil cultivation systems, with 23% of the USA[7] and about half of Argentinian agricultural land under zero-tillage.[98]

This extreme change in soil cultivation drills seeds directly into undisturbed soils, saving considerably on fuel costs but sometimes having negative implications for weed control and yield.[7] A more common approach is the non-inversion tillage of the top 5–15 cm depth of soil, referred to as minimum or conservation tillage. A common description of conservation tillage is that at 30% of surface residue is preserved after seed drilling so that organic matter and structural stability are enhanced, thereby decreasing erosion.[6]

Reduced tillage refers to zero- and conservation-tillage systems. As less soil is disturbed under reduced tillage, shorter-term negative impacts can be an increased soil density and a mechanical impedance to roots at shallower depths, but a positive impact can be a greater pore continuity, so that air permeability and water transport are faster.[99] Over longer periods of time, natural mechanisms improve the soil's functionality for crop production and mechanical stability.[15] Processes that improve soil physical conditions include the creation of root channels, the maintenance of earthworm burrows that are destroyed by intensive cultivation, natural weathering through wetting and drying cycles (intensified by root water uptake) that swell and shrink the soil and frost action that fragments soil through ice lenses.[1]

Reduced tillage is employed to protect agricultural areas from physical degradation such as compaction and erosion.[100] After a shift from conventional (including plough) to reduced or zero tillage, yields often decrease due to decreasing pore volumes and aeration, impaired temperature and a poorer water balance. It must be taken into account that, at the same time, fuel and working time for tillage operations can be saved, so at the farm gate reduced tillage can be more profitable from the outset.[100] After a few years of reduced tillage intensity, the productivity can increase again due to the recovery of soil structure by the mechanisms described above,[101] but not on all soils.[102] Additionally, the soil becomes mechanically more stable,[8] which resists soil compaction by traffic. Along with plant cover and greater organic matter content of surface soils, mechanical stability also provides greater resistance to water and wind erosion.

Reduced tillage is advocated for its potential to sequester carbon in soils, but the capacity in comparison to conventional tillage varies considerably between regions.[103] Increases in soil carbon from the adoption of reduced tillage are greater in the tropics compared with temperate regions because carbon mineralisation by soil cultivation is greater and biomass production is much less. In temperate maritime climates there exists some evidence of no increase in carbon storage, with Sun *et al.*[104] also arguing that biased sampling procedures may limit the reliability of previous data.

A trade off to carbon storage under reduced tillage could be increased emissions of greenhouse gases, particularly nitrous oxide. In a review of 45 sites of data, Rochette[105] concluded that this was only a problem on poorly-aerated soils, with well-drained sites having fewer emissions. Compared to conventional ploughing, fuel use is always much less under reduced tillage, with considerable savings under zero tillage.[100] A shift from conventional to reduced or zero tillage usually increases the amounts of pesticides applied to control weeds. Under zero tillage, increases in weeds by 2–20 times can occur in comparison with ploughed systems.[106] However, shallow, non-inversion cultivation through reduced tillage could be used effectively to help control weed populations whilst also not redistributing the weed seedbank.

4.2 Biological Tillage

Soil compaction can be alleviated by the cultivation of plant species with a strong and deep rooting root system that is able to penetrate mechanically impeded layers. The roots also need to endure high levels of water saturation, poor aeration and low redox potentials for some time. Two plants often used for biological tillage are alfalfa and clover, which have the added benefit of also fixing nitrogen that is vital to soil fertility. The root channels produced under biological tillage produce continuous biopores that can be accessed by subsequent crops. Along these biopores, exudation of organic compounds and water uptake by roots induces aggregation and cracking of soil.[107] Plants with deep and intensive rooting systems also incorporate organic matter into deeper soil depths. Many feel that deep rooting plants offer considerable potential for terrestrial carbon sequestration.[108]

Soil pore structures formed by biological tillage often contain coarse and medium-sized pores. These pores are rapid transmission pathways that improve the flow of water and gases, and therefore promote (re-)colonisation by micro organisms and fauna.[107] The conditions for subsequent crop growth can be greatly improved. For coarse-textured soils, soils which are too wet, or soils with other constraints for subsoiling, biological tillage provides a very good alternative to deep loosening.

4.3 Vegetation and Root Reinforcement

Planting or managing vegetation is a very common practice to combat soil erosion and stabilise unstable slopes against landslides.[109,110] Ugly grey concrete or expensive reinforcing soil nails next to transport corridors are being replaced by a range of plants whose roots reinforce the soil over a range of depths.[111] In agriculture, planting of vegetation in severely degraded soils increases mechanical stability as well as improving soil fertility.[112] The critical factors in their contributions are rooting depth,[113,114] root distribution[115–117] and root diameter.[118]

Much research has focused on the role of woody species in the stabilisation of slopes and embankments due to their ability to penetrate deeper into the soil than plants with a typically fine root system.[119] Tree roots provide both deep anchorage and also reductions in pore water pressure through transpiration.[120,121] As pore water pressures increase, soil instability increases until failure occurs, where roots either break, due to excessive root tensile stresses, or fail through pulling out of the soil. The two failure mechanisms, breakage and pull-out, are generally influenced by two main root properties: root tensile strength and root architecture.[122] The effect of different root architectures within woody root systems has been studied through the use of root analogues to assess pull-out resistance. Tap roots (single roots which grow vertically through the soil) have the least resistance to pull-out whilst dichotomous roots (roots with branches off the main axis) are the most resistant.[122] Woody plant species may be used to increase slope stability in areas prone to mass-wastage events. Smaller scale erosion caused by overland flow of water is managed more effectively by the use of vegetation with fibrous roots.

One of the limitations of woody root systems is that soil-surface coverage is significantly less than that provided by species with a typically fibrous root system, such as grasses. Fibrous root systems have a greater volume of roots in surface soils,[123,124] increasing surface soil stabilisation better than woody roots. Increasing soil coverage reduces splash erosion through interception of rainfall, with plants having a finer root system typically contributing more to surface coverage, and with the combination of canopy and roots reducing soil loss.[125]

Vetiver grass has been researched widely for use in land-restoration projects. Compared with a typical fibrous root system, Vetiver's root system is finer and denser, with the ability to penetrate deeper into the soil. Roots have been found to extend up to 3 metres deep in 12 months with an ability to adapt to adverse soil conditions.[126] The application of such grasses in an agricultural context is to create buffer strips which act to both stabilise soil but also catch eroded material from cropping areas.[126]

The use of vegetation to control erosion and landslides has the added benefit of improving soil fertility. All plants can help restore soil organic matter, critical for maintaining soil structure and soil hydraulic properties. Legumes can also increase soil nitrogen. Soil organic carbon under grass and legumes can increase by *ca.* 20% in 4 years compared to a conventional till wheat system.[127] Restoration of deforested areas in Brazil, using legume trees, showed higher stocks of C and N over a thirteen year period when compared to a deforested area.[128]

Soil erosion can be minimised through the use of vegetation stabilising soils through surface coverage and also root inclusions. Fibrous root systems offer the potential to stabilise the surface soils whilst tree roots penetrate soils deeper and may reduce the risks of mass wastage events such as landslides.

4.4 New Technologies

The recent global adoption of reduced tillage systems was described earlier in this section. There are other technologies that are already in widespread use that aim to decrease the physical damage of soil by farming. Low ground pressure (LGP) tyres allow access to fields over a wider range of weather conditions. Graham *et al.*[129] found that switching to LGP tyres increased wheat yields by 6–7%. Vermeulen and Klooster[130] found a 4% increase in potato yield with LGP tyres. Other studies have found minimal impact.[131] However, the cost of LGP tyres is high so it has been estimated that a farm of at least 200 ha is needed to recoup the costs.[132] Another traction technology in widespread use is tracks. These confine compaction to topsoil and distribute the load more efficiently.[133]

LGP tyres and tracks decrease the contact pressure on soils, which decreases topsoil damage, but subsoil damage depends on the overall weight of the machine.[134] One option is to use smaller tractors, but by nature they require more passes to perform the same task as a larger tractor. With multiple passes, soil damage increases, so lighter tractors can cause as much damage as larger tractors.[135]

A technology attracting considerable attention is Controlled Traffic Farming (CTF). Axle widths of different machinery are matched, with vehicles restricted to dedicated tramlines. The approach was reviewed in detail by Chamen.[10] CTF and standard farming practices are now aided by Global Navigation Satellite Systems (GNSS) that have a spatial accuracy of several centimetres and follow specific tracks within fields. This technology lets farmers map problem areas, apply precision inputs and develop highly efficient cultivation, spraying and harvesting practices, where overlaps between passes are minimised. In a given year, 95% of an agricultural field can experience at least one wheel pass.[136] Using a one-pass cultivation system, such as minimum tillage with auto-steer GNSS, the amount can reduce to 45% and to less than 20% if direct drilling is used.

There is also potential to target improvements to crop roots so that they respond better to physically limited soil conditions. In mechanically impeded soils, for instance, roots preferentially grow through biopores.[84] Screening approaches can select cultivars with root systems that are better able to access biopores.[90] There is a considerable impact of soil physical condition on resource capture between different varieties of the same crop.[89] A major factor driving differences between crop varieties will be the structure of the root system, which has been recognised in the recent shift in plant sciences to consider processes beneath the ground.

4.5 Carbon

Powlson states "*From almost any viewpoint, it is desirable to maintain SOC [soil organic carbon] content at as high a value as possible for the soil type and environment as this is beneficial for a wide range of soil physical properties and root growth*".[137] Various practices exist to return carbon to soils, including the better management of crop stubble, use of cover crops, amendments with animal wastes produced on farm, or imports of carbon from off the farm. New technologies produce forms of carbon, such as biochar, that are retained in soils for long periods of time and are thought to help retain nutrients and pesticides, and build soil structure.[138]

Carbon has various potentially positive impacts on the physical condition of soil for crop production. It is a building block in the formation of stable, aggregated soil[54] that resists erosion and slumping, capping, or hardsetting. Soil compaction damage has also been found to decrease when organic carbon levels are enhanced. With greater carbon, soils are more resilient to compaction, so they bounce back more after wheelings or livestock trampling.[139] Considerable research has been conducted on soil carbon as it underlies so many processes,[140–142] so this chapter has only provided a very brief overview in the context of soil physical conditions.

5 Case Study 1 – Soil Restoration in the Loess Plateau, China

The Loess Plateau in China is unique and covers 624 000 km^2 comprised of very fine soils, making it highly susceptible to wind and water erosion. For over 5000 years the landscape has been influenced by human activity, resulting in widespread degradation.[143] Erosion has been catastrophic due to large-scale overgrazing and overuse of the land, resulting in widespread poverty and abandonment of large areas no longer capable of agricultural productivity. The Loess Plateau contains three main landscapes: tableland and gullies, hilly land and gullies, and sand land. Soil erosion rates are high, with an average 5000–10 000 Mg km^2 per year, with some highly erodible areas producing more than 20 000 Mg km^2 per year.[144]

Natural vegetative restoration has been successful in reducing erosion in all landscape and is widely accepted as the best strategy for minimising soil loss;[145] the type of vegetation, however, is important. In an attempt to increase vegetation cover in erosion-prone areas, the Chinese government launched the 'Grain for Green Project' in 1999. The aim was to increase forest and grassland coverage through the reduction of cropping on steep slopes most prone to erosion. In Shaanxi Province the total vegetation coverage increased by 12.5% in the 7 years since 1998. A large proportion of vegetation increase was attributable to control of grazing by livestock; however, due to the management of afforested sites to increase production, a net decline of 6.1% was found in this type of land use.[146] As previously mentioned in this chapter, trees act to increase soil stability through deep roots and also increase evapotranspiration, reducing pore water pressures in the soil. Within some areas this has caused conflict with increasing water scarcity due to water usage caused by afforestation, necessitating the correct choices for sustainable water management. Trees and shrubs create greater soil moisture depletion in soil depths of 1.0–3.0 m than natural grasslands and may, therefore, not be suitable unless the soils are frequently wet.[147] Other types of vegetation in differing land uses have the ability to increase soil moisture; these include cropland, fallow land, and intercropping land.[148] Pine plantations are widely acknowledged for soil erosion minimisation but may also increase runoff and compound compaction and desiccation of soils, causing soil degradation in semi-arid areas.[149] Due to the complexities of the interactions of plants and soils with the topographical features of the region, structured approaches must be formulated to ensure that benefits to one do not cost further degradation to others.

Slope gradient and rainfall are the significant influencers on soil erosion, with the 'Grain for Green Project' driving changes in vegetation cover as a method to control erosion on steep slopes. Alternative strategies employed have been the terracing of hillsides to enable arable production on slopes previously deemed too steep to cultivate. Not only is soil erosion minimised but soil properties critical for increased yield are improved. Terracing increases soil moisture characteristics, within 1 year of being built increasing moisture content by 8.9–14.45% when compared to sloping land. Soil fertility also increases in terraces, when compared to a 15° slope, with organic matter increasing by 26%, total N 8%, total P 4%, fast-acting N 12% and fast-acting P by 20%. As a result of improvements in soil quality, yield has been shown to increase by 3.6% in the first year after terrace construction and 27.1%, 35.3% and 52.8% after 3, 5 and 7 years, respectively.[150]

Key to understanding the long term-effects of vegetation restoration is having evidence of longer term impacts on soil properties. The Ziwuling Forest within the Loess Plateau covers an area of approximately 23 000 km^2 and contains several areas which have been abandoned for differing lengths of time. Such sites are key in allowing long-term changes in soil properties to be assessed through successional changes over periods of 150 years by natural vegetation. Successional changes were assessed through changes from arable production, to grass, shrubs and finally forest.[151] Soil was densest in the farmland plots (1.29 Mg m^{-3}) and lowest in forested soils (0.99 Mg m^{-3}). Macropore spaces in soil increased over successional changes, implying greater hydraulic conductivity and water-holding capacity. After 14 years of abandonment soil physical properties improved, with an increase in hydraulic conductivity attributable to shrub and tree roots loosening and improving soil structure.[151]

Once soils have become severely eroded a return to a fertile and healthy soil will only occur over extended periods of time. Evidence from the Loess Plateau has shown that one restoration method does not fit all. Stabilisation strategies will help minimise further degradation; however, the methods employed must be chosen with a realistic expectation of the likely outcomes. Arable production of crops from sloping land is clearly unsustainable in environments with historically high erosion rates. In such situations forestry may be better suited, but sustainable management strategies must be applied to ensure soil is protected year-on-year without secondary effects such as water deficits caused by overuse of water by vegetation. Research has demonstrated that several remediation strategies are available and these should be applied relevant to the issue being addressed.

6 Case Study 2 – Recultivated Mine Soils, Eastern Germany (Lusatia)

Germany is the leading lignite producer in the world.[152] Coal mining in Germany is distributed over the three main areas of Rhineland in the west, central Germany and Lusatia in eastern Germany. This chapter concentrates on Lusatia. Germany, with its relatively dense population, needs to reclaim devastated land to ensure agricultural or forest productivity of the areas affected. Returning these areas to agricultural production after mining ceases is referred to as 'recultivation'. Recultivation in Germany, to a wide extent, does not rely on natural succession of the devastated area.[153] The problems of the post-mining areas are varied; natural conditions are modified on the landscape scale. To enable mining, the groundwater has to be lowered below the brown-coal seam (between *ca.* 50 and 400 m below the surface). The impact of ground-water lowering continues for decades after site reclamation, which changes the water balance of whole landscapes. Settlements are moved and finally trees cut. Then the soil material covering the brown-coal deposit is excavated and transported to the other side of the mine with conveyor belts. Excavators are used to spread the substrate as dam-like structures. Thereafter, the recultivation sites are levelled.

Chemical and biological substrate limitations, such as very low or very high pH values[154] and lack of macro and micro nutrients and recent organic matter, can be ameliorated by the application of lime, mineral and/or organic fertilisers, topsoil, and potentially soil microbes. The physical functionality and mechanical stability of the reclamation substrates is generally poor. This is partly because of the substrate properties themselves, but also due to the technical processes of site construction. Substrates excavated from deep in the mine are unstructured and devoid of recent organic carbon.[153,154] The lack of structure and organic carbon in these soils makes them very susceptible to compaction,[154] especially at high water contents. These natural properties are affected by excavating, depositing and levelling the substrates with large-scale heavy machinery, which impart strong mechanical stresses to soils (Figure 5).

Figure 5 Levelling of stockpiled topsoil during the recultivation of a mine in Eastern Germany. The weight of machinery and degradation of the soil in the stockpile presents challenges to physical conditions for crop production if land is returned to agricultural production.

Newly established recultivation sites can have extremely high bulk densities and at the same time comparatively low mechanical stability and very low permeability for gases and liquids.[154] Providing that site construction is conducted very carefully, the quality of reclamation sites under prevailing substrate and climatic conditions can be improved. When machinery is used to minimise small height differences of the surface soil, the machinery that performs the levelling needs to be as light as possible.

The very low nutrient contents, biological activity, and mechanical stability of mine soils in the Lusatia region make them far more susceptible to compaction than natural soils. Even restoration practices such as soil tillage operations need to be minimised to avoid further compaction and physical damage of the soil.[152] However, deep ploughing at the beginning of reclamation and again after approximately 3 years is still recommended, although the lasting effect of such measures is mostly low, especially in coarse substrates. If mined sites are to be returned to agricultural production, crop rotations are applied during site recultivation to improve soil structure and soil organic matter content. Crops used for this purpose should utilise the substrate, require minimal nutrients and preferably fix nitrogen (*e.g.* peas, *Pisum*; lupines, *Lupinus*; field beans, *Vicia*), use water efficiently (*e.g. Melilotus*, sweet clover or *Gramineae*, grass mixes), develop deep root systems (*e.g.* alfalfa) and produce large amounts of above- and below-ground biomass to be incorporated into the infertile soil substrates that cover the surface of the mine.

Reclamation in Lusatia is still receiving considerable research. Reclamation in most cases ensures approximately the same productivity of agricultural crops as before within a few decades,[152] at least according to German legislation. However, the time needed to fully recultivate mined soils to agricultural production remains contentious. Nevertheless, the practice of rebuilding agriculture on an infertile and physically sensitive substrate provides valuable information for the restoration of all soils.

7 Conclusions

Food security relies on good soil physical condition. It is therefore worrying that about 24% of the global inhabitable land area is affected by soil degradation, with erosion being the cause of at least half of this damage. Soil compaction is probably a much greater problem than is currently realised, as the evidence-base focuses on only a few regions. Trends such as increasing machinery weight suggest worsening impacts. Intensive cultivation has depleted soils of carbon, making them more vulnerable to physical damage.

There are many opportunities, however, to improve soil physical conditions to help address concerns about food security and also environmental protection. A shift in farming towards reduced tillage systems provides benefits to many regions, with the Cerrado area of Brazil demonstrating a huge impact where almost unviable land was made very productive. Engineering technologies such as satellite navigation for auto-steer tractors, Controlled

Traffic Farming and precision agriculture are other approaches to avoid or mitigate soil physical damage.

The impact of soil physical conditions on crop production is well understood. There is pressing need for soil and plant scientists to work together to develop the next generation of crops that perform better when soil conditions are not ideal. Deep roots that capture nutrients and water stored in soil are an essential component of sustainable agriculture.

References

1. P. D. Hallett and A. G. Bengough, in *Russell's Soil Conditions and Plant Growth*, ed. P. J. Gregory and S. Nortcliff, Wiley-Blackwell, Oxford, in press.
2. L. M. Bresson, *Aust. J. Soil Res.*, 1995, **33**, 195.
3. K. Y. Chan, D. P. Heenan and R. Ashley, *Soil Till. Res.*, 1994, **28**, 301.
4. A. J. Franzluebbers, *Soil Till. Res.*, 2007, **94**, 265.
5. B. R. Warkentin, *Advan. Agron.*, 2008, **97**, 239.
6. N. L. Morris, P. C. H. Miller, J. H. Orson and R. J. Froud-Williams, *Soil Till. Res.*, 2010, **108**, 1.
7. G. B. Triplett and W. A. Dick, *Agron. J.*, 2008, **100**, S153.
8. R. Horn, H. Taubner, M. Wuttke and T. Baumgartl, *Soil Till. Res.*, 1994, **30**, 187.
9. A. P. Whitmore and W. R. Whalley, *J. Exp. Bot.*, 2009, **60**, 2845.
10. W. C. T. Chamen, PhD thesis, School of Applied Sciences, Cranfield University, 2011.
11. S. A. Materechera, *Soil Till. Res.*, 2009, **105**, 251.
12. D. Or and T. A. Ghezzehei, *Soil Till. Res.*, 2002, **64**, 41.
13. D. R. Montgomery, *Proc. Natl. Acad. Sci. U. S. A.*, 2007, **104**, 13268.
14. R. E. Phillips, R. L. Blevins, G. W. Thomas, W. W. Frye and S. H. Phillips, *Science*, 1980, **208**, 1108.
15. J. Roger-Estrade, C. Anger, M. Bertrand and G. Richard, *Soil Till. Res.*, 2010, **111**, 33.
16. The World Food Prize Foundation, 2006 Laureates; www.worldfoodprize.org/en/laureates/20002009_laureates/2006_lobato_mcclung_paolinelli, 2006, accessed 19/06/2012.
17. A. Zink, H. Fleige and R. Horn, *Soil Till. Res.*, 2011, **114**, 127.
18. A. C. Bailey, R. L. Raper, C. E. Johnson and E. C. Burt, *J. Agric. Eng. Res.*, 1995, **61**, 73.
19. T. Keller and J. Arvidsson, *Eur. J. Soil Sci.*, 2007, **58**, 1373.
20. M. F. O'Sullivan, J. K. Henshall and J. W. Dickson, *Soil Till. Res.*, 1999, **49**, 325.
21. B. Zhang, R. Horn and P. D. Hallett, *Soil Sci. Soc. Am. J.*, 2005, **69**, 864.
22. R. Horn and H. Fleige, *Soil Till. Res.*, 2009, **102**, 201.
23. C. W. Watts and A. R. Dexter, *Proceedings of the 14th International Society of Tillage Research Conference*, Pulawy, Poland, 1997, 669.

24. A. S. Gregory, C. W. Watts, W. R. Whalley, H. L. Kuan, B. S. Griffiths, P. D. Hallett and A. P. Whitmore, *Eur. J. Soil Sci.*, 2007, **58**, 1221.
25. G. Cooper, M. B. McGechan and A. J. A. Vinten, *J. Agric. Eng. Res.*, 1997, **68**, 253.
26. K. Hart and D. Baldock, *Greening the CAP: Delivering Environmental Outcomes through Pillar One*, Institute of Environmental Policy, 2011.
27. P. Defossez and G. Richard, *Soil Till. Res.*, 2002, **67**, 41.
28. J. J. H. van den Akker, J. Arvidsson and R. Horn, *Soil Till. Res.*, 2003, **73**, 1.
29. P. Barré, B. M. McKenzie and P. D. Hallett, *Soil Biol. Biochem.*, 2009, **41**, 656.
30. R. Horn, H. Fleige, F. H. Richter, E. A. Czyz, A. Dexter, E. Diaz-Pereira, E. Dumitru, R. Enarche, F. Mayol, K. Rajkai, D. de la Rosa and C. Simota, *Soil Till. Res.*, 2005, **82**, 47.
31. R. Horn, J. Vossbrink, S. Peth and S. Becker, *For. Ecol. Manage.*, 2007, **248**, 56.
32. L. Pietola, R. Horn and M. Yli-Halla, *Soil Till. Res.*, 2005, **82**, 99.
33. I. M. Young, E. Blanchart, C. Chenu, M. Dangerfield, C. Fragoso, M. Grimaldi and J. J. M. L. Ingram, *Global Change Biol.*, 1998, 4.
34. R. L. Schafer, C. E. Johnson, A. J. Koolen, S. C. Gupta and R. Horn, *Trans. ASAE*, 1992, **35**, 1761.
35. L. R. Oldeman, R. T. A. Hakkeling and W. G. Sombroek, *World Map of the Status of Human-induced Soil Degradation. An Explanatory Note*, ISRIC, Wageningen, The Netherlands/UNEP: Nairobi, Kenya. 1991.
36. B. C. Ball, D. J. Campbell and E. A. Hunter, *Soil Till. Res.*, 2000, **57**, 83.
37. B. Soane, *Overcompaction of Soils on Scottish Farms: A Survey*, Scottish Institute of Agricultural Engineering, 1987, pp. 18.
38. B. D. Soane and C. Vanouwerkerk, *Soil Till. Res.*, 1995, **35**, 5.
39. H. J. Koch, J. Dieckmann, A. Büchse and B. Märländer, *Eur. J. Agron.*, 2009, **30**, 101.
40. W. R. Whalley, P. B. Leeds-Harrison, L. J. Clark and D. J. G. Gowing, *Soil Till. Res.*, 2005, **84**, 18.
41. M. Silgram, D. R. Jackson, A. Bailey, J. Quinton and C. Stevens, *Earth Surface Processes Landforms*, 2010, **35**, 699.
42. A. Beylich, H. R. Oberholzer, S. Schrader, H. Hoper and B. M. Wilke, *Soil Till. Res.*, 2010, **109**, 133.
43. C. J. Shestak and M. D. Busse, *Soil Sci. Soc. Am. J.*, 2005, **69**, 236.
44. H. D. Kutzbach, *J. Agric. Eng. Res.*, 2000, **76**, 237.
45. P. Schjönning, G. Heckrath and B. Christensen, *Threats to Soil Quality in Denmark: A Review of Existing Knowledge in the Context of the EU Soil Thematic Strategy*, DJF Rapport, Markbrug, 2009, 143.
46. T. Batey, *Soil Use Manage.*, 2009, **25**, 335.
47. A. Knapen, J. Poesen, G. Govers, G. Gyssels and J. Nachtergaele, *Earth Sci. Rev.*, 2007, **80**, 75.
48. Y. Le Bissonnais, *Eur. J. Soil Sci.*, 1996, **47**, 425.

49. J. H. Zhang, S. J. Ni and Z. A. Su, *Eur. J. Soil Sci.*, 2012, **63**, 165.
50. D. Pimentel, *Environ. Dev. Sustain.*, 2006, **8**, 119.
51. J. M. Tisdall and J. M. Oades, *Aust. J. Soil Res.*, 1979, **17**, 429.
52. P. Barré and P. D. Hallett, *Eur. J. Soil Sci.*, 2009, **60**, 525.
53. S. Czarnes, P. D. Hallett, A. G. Bengough and I. M. Young, *Eur. J. Soil Sci.*, 2000, **51**, 435.
54. J. Six, K. Paustian, E. T. Elliott and C. Combrink, *Soil Sci. Soc. Am. J.*, 2000, **64**, 681.
55. R. Wright, *A Short History of Progress*, House of Anansi Press, Toronto, 2004, pp. 211.
56. F. G. A. Verheijen, R. J. A. Jones, R. J. Rickson and C. J. Smith, *Earth Sci. Rev.*, 2009, **94**, 23.
57. F. F. Pruski and M. A. Nearing, *Water Resour. Res.*, 2002, 38.
58. D. Mullan, D. Favis-Mortlock and R. Fealy, *Agr. For. Meteorol.*, 2012, **156**, 18.
59. M. L. Clarke and H. M. Rendell, *Earth Surf. Processes Landforms*, 2010, **35**, 1281.
60. U. C. Sharma and V. Sharma, in *Risk in Water Resources Management*, ed. G. Bloschl, K Takeuchi, S Farnleitner and A Schumann, IAHS Publication, Wallingford, 2011, 253–258.
61. AGRA, *Driving Real Change. AGRA in 2010*, Alliance for a Green Revolution in Africa (AGRA), Nairobi, Kenya, 2011.
62. J. R. Finney and B. A. G. Knight, J. *Agric. Sci.*, 1973, **80**, 435.
63. T. A. Ghezzehei and D. Or, *Soil Sci. Soc. Am. J.*, 2003, **67**, 12.
64. S. Assouline and Y. Mualem, *Water Resour. Res.*, 2000, **36**, 2341.
65. M. Diacono and F. Montemurro, *Agron. Sustainable Dev.*, 2010, **30**, 401.
66. J. Six, H. Bossuyt, S. Degryze and K. Denef, *Soil Till. Res.*, 2004, **79**, 7.
67. W. W. Emerson, *Nature*, 1959, **183**, 538.
68. C. Mullins, D. MacLeod, K. Northcote, J. Tisdall and I. Young, *Adv. Soil Sci.*, 1990, **11**, 37.
69. K. Y. Chan, D. P. Heenan and H. B. So, *Aust. J. Exp. Agric.*, 2003, **43**, 325.
70. D. Or, F. J. Leij, V. A. Snyder and T. A. Ghessehei, *Water Resour. Res.*, 2000, **36**, 1641.
71. A. G. Bengough, B. M. McKenzie, P. D. Hallett and T. A. Valentine, *J. Exp. Bot.*, 2011, **62**, 59.
72. A. R. Dexter, *Soil Till. Res.*, 1988, **11**, 199.
73. W. D. Reynolds, B. T. Bowman, C. F. Drury, C. S. Tan and X. Lu, *Geoderma*, 2002, **110**, 131.
74. M. Tapela and T. S. Colvin, *Soil Till. Res.*, 2002, **64**, 203.
75. W. D. Reynolds, C. F. Drury, C. S. Tan, C. A. Fox and X. M. Yang, *Geoderma*, 2009, **152**, 252.
76. E. A. Czyz, *Soil Till. Res.*, 2004, **79**, 153.
77. A. R. Dexter, *Geoderma*, 2004, **120**, 201.
78. A. R. Dexter, *Geoderma*, 2004, **120**, 215.

79. A. R. Dexter, E. A. Czyz, M. Birkas, E. Diaz-Pereira, E. Dumitru, R. Enache, H. Fleige, R. Horn, K. Rajkaj, D. de la Rosa and C. Simota, *Soil Till. Res.*, 2005, **82**, 29.
80. L. L. Li, K. Y. Chan, Y. N. Niu, G. D. Li, A. Oates, A. Dexter and G. B. Huang, *Soil Till. Res.*, 2011, **113**, 82.
81. A. N. Beutler, J. F. Centurion and A. P. d. Silva, *Braz. Arch. Biol. Technol.*, 2005, **48**, 863.
82. A. P. da Silva and B. D. Kay, *Plant Soil*, 1996, **184**, 323.
83. W. Ehlers, U. Kopke, F. Hesse and W. Bohm, *Soil Till. Res.*, 1983, **3**, 261.
84. R. G. White and J. A. Kirkegaard, *Plant Cell Environ.*, 2010, **33**, 133.
85. J. A. Kirkegaard and J. M. Lilley, *Aust. J. Exp. Agric.*, 2007, **47**, 590.
86. T. A. Valentine, P. D. Hallett, K. Binnie, M. Young, G. R. Squire, C. Hawes and A. G. Bengough, *Ann. Bot.*, 2012, **110**, 259.
87. J. A. Kirkegaard and J. R. Hunt, *J. Exp. Bot.*, 2010, **61**, 4129.
88. A. C. Newton, D. C. Guy, A. G. Bengough, D. C. Gordon, B. M. McKenzie, B. Sun, T. A. Valentine and P. D. Hallett, *Field Crops Res.*, 2012, **128**, 91.
89. T. S. George, L. K. Brown, A. C. Newton, P. D. Hallett, B. H. Sun, W. T. B. Thomas and P. J. White, *Plant Soil*, 2011, **339**, 113.
90. B. M. McKenzie, A. G. Bengough, P. D. Hallett, W. T. B. Thomas, B. Forster and J. W. McNicol, *Field Crops Res.*, 2009, **112**, 165.
91. M. A. J. Parry, M. Reynolds, M. E. Salvucci, C. Raines, P. J. Andralojc, X. G. Zhu, G. D. Price, A. G. Condon and R. T. Furbank, *J. Exp. Bot.*, 2010.
92. P. Savadogo, L. Sawadogo and D. Tiveau, *Agric. Ecosyst. Environ.*, 2007, **118**, 80.
93. J. C. Medeiros, A. P. da Silva, C. E. Pellegrino Cerri, N. F. Balarezo Giarola, G. C. Figueiredo and F. J. Cury Fracetto, *Plant Soil*, 2011, **338**, 5.
94. N. R. Hulugalle and F. Scott, *Aust. J. Soil Res.*, 2008, **46**, 173.
95. M. J. Marks and G. C. Soane, *Soil Use Manage.*, 1987, **3**, 115.
96. G. Spoor, F. G. J. Tijink and P. Weisskopf, *Soil Till. Res.*, 2003, **73**, 175.
97. K. Hartge, *Soil Till. Res.*, 1980, **1**, 187.
98. P. K. Farage, J. Ardo, L. Olsson, E. A. Rienzi, A. S. Ball and J. N. Pretty, *Soil Till. Res.*, 2007, **94**, 457.
99. R. A. Comia, M. Stenberg, P. Nelson, T. Rydberg and I. Håkansson, *Soil Till. Res.*, 1994, **29**, 335.
100. B. D. Soane, B. C. Ball, J. Arvidsson, G. Basch, F. Moreno and J. Roger-Estrade, *Soil Till. Res.*, 2012, **118**, 66.
101. M. W. Strudley, T. R. Green and J. C. Ascough, *Soil Till. Res.*, 2008, **99**, 4.
102. X. H. Shi, X. M. Yang, C. F. Drury, W. D. Reynolds, N. B. McLaughlin, T. W. Welacky and X. P. Zhang, *Soil Sci. Soc. Am. J.*, 2011, **75**, 1083.
103. S. M. Ogle, F. J. Breidt and K. Paustian, *Biogeochemistry*, 2005, **72**, 87.
104. B. H. Sun, P. D. Hallett, S. Caul, T. J. Daniell and D. W. Hopkins, *Plant Soil*, 2011, **338**, 17.
105. P. Rochette, *Soil Till. Res.*, 2008, **101**, 97.

106. S. Gruber, C. Pekrun, J. Mohring and W. Claupein, *Soil Till. Res.*, 2012, **121**, 49.
107. T. Kautz, C. Stumm, R. Kösters and U. Köpke, *J. Plant Nutr. Soil Sci.*, 2010, **173**, 490.
108. D. B. Kell, *Ann. Bot.*, 2011, **108**, 407.
109. S. De Baets, D. Torri, J. Poesen, M. P. Salvador and J. Meersmans, *Earth Surface Processes Landforms*, 2008, **33**, 1948.
110. L. J. Waldron and S. Dakessian, *Soil Science*, 1981, **132**, 427.
111. D. H. Gray, *J. Geotech. Eng. ASCE*, 1992, **118**, 1395.
112. X. Peng, B. Zhang, Q. Zhao, R. Horn and P. D. Hallett, *Geoderma*, 2003, **115**, 313.
113. B. B. Docker and T. C. T. Hubble, *Geomorphology*, 2008, **100**, 401.
114. G. B. Bischetti, E. A. Chiaradia, T. Epis and E. Morlotti, *Plant Soil*, 2009, **324**, 71.
115. S. De Baets, J. Poesen, B. Reubens, K. Wemans, J. De Baerdemaeker and B. Muys, *Plant Soil*, 2008, **305**, 207.
116. B. Reubens, J. Poesen, F. d. r. Danjon, G. Geudens and B. Muys, *Trees – Struct. Funct.*, 2007, **21**, 385.
117. N. Osman and S. S. Barakbah, *Ecol. Eng.*, 2006, **28**, 90.
118. A. Stokes, C. Atger, A. G. Bengough, T. Fourcaud and R. C. Sidle, *Plant Soil*, 2009, **324**, 1.
119. T. M. Wynn, S. Mostaghimi, J. A. Burger, A. A. Harpold, M. B. Henderson and L. A. Henry, *J. Environ. Qual.*, 2004, **33**, 2030.
120. C. C. Fan and C. F. Su, *Ecol. Eng.*, 2008, **33**, 157.
121. N. S. Nilaweera and P. Nutalaya, *Bull. Eng. Geol. Environ.*, 1999, **57**, 337.
122. S. B. Mickovski, A. G. Bengough, M. F. Bransby, M. C. R. Davies, P. D. Hallett and R. Sonnenberg, *Eur. J. Soil Sci.*, 2007, **58**, 1471.
123. J. Canadell, R. B. Jackson, J. R. Ehleringer, H. A. Mooney, O. E. Sala and E. D. Schulze, *Oecologia*, 1996, **108**, 583.
124. R. B. Jackson, J. Canadell, J. R. Ehleringer, H. A. Mooney, O. E. Sala and E. D. Schulze, *Oecologia*, 1996, **108**, 389.
125. G. Gyssels, J. Poesen, E. Bochet and Y. Li, *Prog. Phys. Geogr.*, 2005, **29**, 189.
126. P. A. Dalton, R. J. Smith and P. N. V. Truong, *Agr. Water Manage.*, 1996, **31**, 91.
127. R. C. Dalal, W. M. Strong, E. J. Weston, J. E. Cooper, K. J. Lehane, A. J. King and C. J. Chicken, *Aust. J. Exp. Agric.*, 1995, **35**, 903.
128. M. Macedo, A. Resende, P. Garcia, R. Boddey, C. Jantalia, S. Urquiaga, E. Campello and A. Franco, *For. Ecol. Manage.*, 2008, **255**, 1516.
129. J. P. Graham, P. S. Blackwell, J. V. Armstrong, D. G. Christian, K. R. Howse, C. J. Dawson and A. R. Butler, *Soil Till. Res.*, 1986, **7**, 189.
130. G. D. Vermeulen and J. J. Klooster, *Soil Till. Res.*, 1992, **24**, 337.
131. W. C. T. Chamen, E. T. Chittey, P. R. Leede, M. J. Goss and K. R. Howse, J. *Agric. Eng. Res.*, 1990, **47**, 1.

132. K. Eradat Oskoui, D. J. Campbell, B. D. Soane and M. J. McGregor, in *Soil Compaction in Crop Production*, ed. B. D. Soane and C. van Ouwerkerk, Elsevier, Amsterdam, 1994.
133. M. Pagliai, A. Marsili, P. Servadio, N. Vignozzi and S. Pellegrini, *Soil Till. Res.*, 2003, **73**, 119.
134. M. Lamandé and P. Schjönning, *Soil Till. Res.*, 2011, **114**, 71.
135. D. Jorajuria, L. Draghi and A. Aragon, *Soil Till. Res.*, 1997, **41**, 1.
136. M. Kroulík, F. Kumhála, J. Hůla and I. Honzík, *Soil Till. Res.*, 2009, **105**, 171.
137. D. S. Powlson, P. J. Gregory, W. R. Whalley, J. N. Quinton, D. W. Hopkins, A. P. Whitmore, P. R. Hirsch and K. W. T. Goulding, *Food Policy*, 2011, **36**, S72.
138. R. S. Kookana, A. K. Sarmah, L. Van Zwieten, E. Krull and B. Singh, *Adv. Agron.*, 2011, **112**, 103.
139. H. L. Kuan, P. D. Hallett, B. S. Griffiths, A. S. Gregory, C. W. Watts and A. P. Whitmore, *Eur. J. Soil Sci.*, 2007, **58**, 811.
140. H. H. Janzen, *Agric. Ecosyst. Environ.*, 2004, **104**, 399.
141. R. Lal, *Science*, 2004, **304**, 1623.
142. P. Loveland and J. Webb, *Soil Till. Res.*, 2003, **70**, 1.
143. H. Shi and M. G. Shao, *J. Arid Environ.*, 2000, **45**, 9.
144. L. D. Chen, W. Wei, B. J. Fu and Y. H. Lu, *Prog. Phys. Geogr.*, 2007, **31**, 389.
145. F. L. Zheng, *Pedosphere*, 2006, **16**, 420.
146. S. Cao, L. Chen and X. Yu, *J. Appl. Ecol.*, 2009, **46**, 536.
147. Y. L. Huang, L. D. Chen, B. J. Fu, L. P. Zhang and Y. L. Wang, *Pedosphere*, 2005, **15**, 509.
148. B. J. Fu, J. Wang, L. D. Chen and Y. Qiu, *Catena*, 2003, **54**, 197.
149. L. Chen, J. Wang, W. Wei, B. Fu and D. Wu, *For. Ecol. Manage.*, 2010, **259**, 1291.
150. X. H. Liu, B. L. He, Z. X. Li, J. L. Zhang, L. Wang and Z. Wang, *Environ. Earth Sci.*, 2011, **62**, 797.
151. Y. Y. Li and M. A. Shao, *J. Arid Environ.*, 2006, **64**, 77.
152. J. Krümmelbein, O. Bens, T. Raab and M. Anne Naeth, *Can. J. Soil. Sci.*, 2012, **92**, 53.
153. F. Schulz, *Drei Jahrhunderte Lausitzer Braunkohlebergbau*, Lusatia Verlag, Bautzen, 2000.
154. J. Krümmelbein, R. Horn, T. Raab, O. Bens and R. F. Hüttl, *Soil Till. Res.*, 2010, **111**, 19.

Subject Index